Reviews of Environmental Contamination and Toxicology

Editor
George W. Ware

Associate Editor
David M. Whitacre

Founding Editor
Francis A. Gunther

VOLUME 191

Springer

Coordinating Board of Editors

Springer
New York: 233 Spring Street, New York, NY 10013, USA
Heidelberg: Postfach 10 52 80, 69042 Heidelberg, Germany

ISBN 978-1-4419-2403-2 e-ISBN 978-0-387-69163-3

ISSN 0179-5953

Printed on acid-free paper.

springer.com

Reviews of Environmental Contamination and Toxicology

VOLUME 191

Foreword

International concern in scientific, industrial, and governmental communities over traces of xenobiotics in foods and in both abiotic and biotic environments has justified the present triumvirate of specialized publications in this field: comprehensive reviews, rapidly published research papers and progress reports, and archival documentations. These three international publications are integrated and scheduled to provide the coherency essential for nonduplicative and current progress in a field as dynamic and complex as environmental contamination and toxicology. This series is reserved exclusively for the diversified literature on "toxic" chemicals in our food, our feeds, our homes, recreational and working surroundings, our domestic animals, our wildlife and ourselves. Tremendous efforts worldwide have been mobilized to evaluate the nature, presence, magnitude, fate, and toxicology of the chemicals loosed upon the earth. Among the sequelae of this broad new emphasis is an undeniable need for an articulated set of authoritative publications, where one can find the latest important world literature produced by these emerging areas of science together with documentation of pertinent ancillary legislation.

Research directors and legislative or administrative advisers do not have the time to scan the escalating number of technical publications that may contain articles important to current responsibility. Rather, these individuals need the background provided by detailed reviews and the assurance that the latest information is made available to them, all with minimal literature searching. Similarly, the scientist assigned or attracted to a new problem is required to glean all literature pertinent to the task, to publish new developments or important new experimental details quickly, to inform others of findings that might alter their own efforts, and eventually to publish all his/her supporting data and conclusions for archival purposes.

In the fields of environmental contamination and toxicology, the sum of these concerns and responsibilities is decisively addressed by the uniform, encompassing, and timely publication format of the Springer triumvirate:

Reviews of Environmental Contamination and Toxicology [Vol. 1 through 97 (1962–1986) as Residue Reviews] for detailed review articles concerned with any aspects of chemical contaminants, including pesticides, in the total environment with toxicological considerations and consequences.

Bulletin of Environmental Contamination and Toxicology (Vol. 1 in 1966) for rapid publication of short reports of significant advances and discoveries in the fields of air, soil, water, and food contamination and pollution as well as methodology and other disciplines concerned with the introduction, presence, and effects of toxicants in the total environment.

Archives of Environmental Contamination and Toxicology (Vol. 1 in 1973) for important complete articles emphasizing and describing original experimental or theoretical research work pertaining to the scientific aspects of chemical contaminants in the environment.

Manuscripts for *Reviews* and the *Archives* are in identical formats and are peer reviewed by scientists in the field for adequacy and value; manuscripts for the *Bulletin* are also reviewed, but are published by photo-offset from camera-ready copy to provide the latest results with minimum delay. The individual editors of these three publications comprise the joint Coordinating Board of Editors with referral within the Board of manuscripts submitted to one publication but deemed by major emphasis or length more suitable for one of the others.

Coordinating Board of Editors

Preface

The role of *Reviews* is to publish detailed scientific review articles on all aspects of environmental contamination and associated toxicological consequences. Such articles facilitate the often-complex task of accessing and interpreting cogent scientific data within the confines of one or more closely related research fields.

In the nearly 50 years since *Reviews of Environmental Contamination and Toxicology* (formerly *Residue Reviews*) was first published, the number, scope and complexity of environmental pollution incidents have grown unabated. During this entire period, the emphasis has been on publishing articles that address the presence and toxicity of environmental contaminants. New research is published each year on a myriad of environmental pollution issues facing peoples worldwide. This fact, and the routine discovery and reporting of new environmental contamination cases, creates an increasingly important function for *Reviews*.

The staggering volume of scientific literature demands remedy by which data can be synthesized and made available to readers in an abridged form. *Reviews* addresses this need and provides detailed reviews worldwide to key scientists and science or policy administrators, whether employed by government, universities or the private sector.

There is a panoply of environmental issues and concerns on which many scientists have focused their research in past years. The scope of this list is quite broad, encompassing environmental events globally that affect marine and terrestrial ecosystems; biotic and abiotic environments; impacts on plants, humans and wildlife; and pollutants, both chemical and radioactive; as well as the ravages of environmental disease in virtually all environmental media (soil, water, air). New or enhanced safety and environmental concerns have emerged in the last decade to be added to incidents covered by the media, studied by scientists, and addressed by governmental and private institutions. Among these are events so striking that they are creating a paradigm shift. Two in particular are at the center of ever-increasing media as well as scientific attention: bioterrorism and global warming. Unfortunately, these very worrisome issues are now super-imposed on the already extensive list of ongoing environmental challenges.

The ultimate role of publishing scientific research is to enhance understanding of the environment in ways that allow the public to be better informed, The term "informed public" as used by Thomas Jefferson in the age of enlightenment conveyed the thought of soundness and good judgment. In the modern sense, being "well informed" has the narrower meaning of having access to sufficient information. Because the public still gets most of its information on science and technology from TV news and reports, the role for scientists as interpreters and brokers of scientific information to the public will grow rather than diminish.

Environmentalism is the newest global political force, resulting in the emergence of multi-national consortia to control pollution and the evolution of the environmental ethic. Will the new politics of the 21st century involve a consortium of technologists and environmentalists, or a progressive confrontation? These matters are of genuine concern to governmental agencies and legislative bodies around the world.

For those who make the decisions about how our planet is managed, there is an ongoing need for continual surveillance and intelligent controls, to avoid endangering the environment, public health, and wildlife. Ensuring safety-in-use of the many chemicals involved in our highly industrialized culture is a dynamic challenge, for the old, established materials are continually being displaced by newly developed molecules more acceptable to federal and state regulatory agencies, public health officials, and environmentalists.

Reviews publishes synoptic articles designed to treat the presence, fate, and, if possible, the safety of xenobiotics in any segment of the environment. These reviews can either be general or specific, but properly lie in the domains of analytical chemistry and its methodology, biochemistry, human and animal medicine, legislation, pharmacology, physiology, toxicology and regulation. Certain affairs in food technology concerned specifically with pesticide and other food-additive problems may also be appropriate.

Because manuscripts are published in the order in which they are received in final form, it may seem that some important aspects have been neglected at times. However, these apparent omissions are recognized, and pertinent manuscripts are likely in preparation or planned. The field is so very large and the interests in it are so varied that the Editor and the Editorial Board earnestly solicit authors and suggestions of underrepresented topics to make this international book series yet more useful and worthwhile.

Justification for the preparation of any review for this book series is that it deals with some aspect of the many real problems arising from the presence of foreign chemicals in our surroundings. Thus, manuscripts may encompass case studies from any country. Food additives, including pesticides, or their metabolites that may persist into human food and animal feeds are within this scope. Additionally, chemical contamination in any manner of air, water, soil, or plant or animal life is within these objectives and their purview.

Manuscripts are often contributed by invitation. However, nominations for new topics or topics in areas that are rapidly advancing are welcome. Preliminary communication with the Editor is recommended before volunteered review manuscripts are submitted.

Tucson, Arizona G.W.W.

Table of Contents

Rev Environ Contam Toxicol 191:1–22

Polycyclic Aromatic Hydrocarbons in the South American Environment

Ricardo Barra, Caroline Castillo, and Joao Paulo Machado Torres

Contents

I. Introduction

Polycyclic aromatic hydrocarbons (PAHs) are a large group of chemical substances with a similar structure comprising two or more joined aromatic carbon rings. PAHs vary both in their chemical characteristics and in their environmental sources, and they are found in the environment both as gases and associated with particulate material.

These substances are widespread contaminants throughout the environment, arising from both anthropogenic and natural sources such as fossil fuel combustion, the direct release of oil and oil products, and uncontrolled combustion processes. They are a global issue because they can be transported over long distances through the atmosphere, and they have been intensively studied for reasons of their mutagenic and carcinogenic properties.

Communicated by Dr. Lilia Albert.

R. Barra (✉, C. Castillo)
Aquatic Systems Research Unit, EULA-Chile Environmental Sciences Center, Barrio Universitario S/N, University of Concepción, P.O. Box 160-C, Concepción, Chile

J.P.M. Torres
Laboratório de Radioisótopos Eduardo Penna Franca, Instituto de Biofísica Carlos Chagas Filho, Universidade Federal do Rio de Janeiro, Ilha do Fundao, Rio de Janeiro Brazil

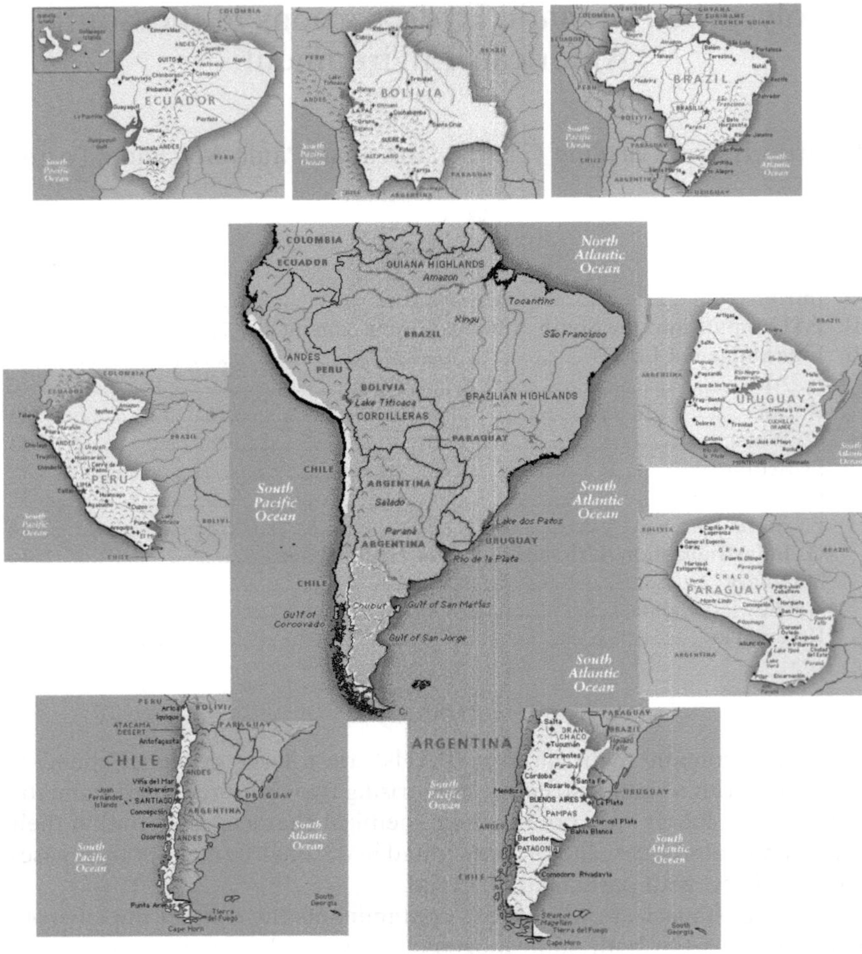

Fig. 1. South American Countries considered in this review.

Some PAHs are classified as priority pollutants by both the U.S. Environmental Protection Agency (USEPA) and the European Union; 16 compounds are designated of environmental interest and 6 of these are known as possible or probable human carcinogens (benzo[*a*]anthracene, benzo[*b*]fluoranthene, benzo[*k*]fluoranthene, benzo[*a*]pyrene, dibenzo[*ah*]-anthracene, and indeno[1,2,3-*cd*]pyrene (De Martinis et al. 2002).

In developing countries, little attention has been paid to PAH pollution, even when their emission sources could be greater than in developed countries. Indeed, Hafner et al. in 2005 concluded that PAH levels in atmospheric samples from developing countries had higher concentrations than other sites in a global survey. The aim of the present review is to provide data on PAH distribution in South American countries as part of a global assessment (Fig. 1). Human impact is documented by examples of PAH pollution in practically every environmental compartment, and local PAH distribution patterns in urban environments are also discussed.

This review presents data summaries collected from a variety of sources, in particular, international peer-reviewed journals, governmental reports, and finally contributions from individual researchers of the different countries belonging to the region.

II. PAH Inputs into the South American Environment

PAHs are unintended by-products that are principally derived from natural and anthropogenic combustion sources (motor vehicle exhaust, wood and coal heating, certain industrial operations, and aluminium plants). In the environment, because of their low water solubility, PAHs are concentrated in bottom sediments and biota where they exert well-characterized toxic effects (i.e., carcinogenic, mutagenic). They are quickly metabolized by fish, although not by invertebrates (Toro et al. 2003).

The air compartment is one of the principal transport pathways in the environment. Atmospheric PAHs are partitioned between two phases, gas and particle; this partition depends strongly on PAH molecular weight. At low molecular weights, PAHs present/display discharge concentrations in the vapor phase, whereas at high molecular weights they are associated with particle material; among these are most of the carcinogenic PAHs. For example, 2- to 3-ring PAHs of low molecular weight are typically found in the gaseous phase while PAHs with 4–6 rings are found mostly in the particulate phase (PM10 and PM2.5) (Sienra et al. 2002).

The extensive forest resources and their intensive use (intentional and unintentional biomass burning) make PAHs one of the high-priority pollutants for the region. Petrogenic PAH sources are also significant, mainly related to petroleum exploitation, refining, and transport. The PAH emission estimates calculated with USEPA 7-PAHs and 16-PAHs emission factors indicate a total emission of 111–500 t polycyclic organic matter/yr for Argentina, principally derived from wildfires (49%), residential wood combustion (31%), and coke production (8%). In Brazil, emissions were estimated at 467–6,607 t/yr, with wood combustion accounting for at least 90% of these values (UNEP 2002).

III. Environmental Levels

A. Air

Most of the data reviewed came from the analysis of atmospheric samples in urban environments. For South America, the most frequently analyzed

PAHs in air samples have been those in Southern Patagonia and in the city of La Plata in Argentina, reporting four datasets corresponding to PAHs in particulate matter in Southern Patagonia. Particulate PAH concentrations in air from Puerto Madryn averaged 6 ng/m^3, whereas in a local industrial park, the site of the country's largest aluminum plant, values reached 1,000 ng/m^3 (Ares and Zavatti 1993). Benzo[*a*]pyrene predominated in stack stream samples, but decreased markedly in air samples, which was attributed to its faster decay relative to benzo[*ghi*]perylene and especially to the more abundant benzo[*k*]fluoranthene. In the city of La Plata (Buenos Aires, Argentina), the total particulate PAHs ranged from 3 to 30 ng/m^3, and the highest concentrations corresponded to fall and winter because of reduction of photochemical activity during the cold months (Catoggio et al. 1989). Pyrene, phenanthrene, benzo[*a*]pyrene, and benzo[*a*]anthracene predominated, indicating the importance of pyrogenic sources. Benzo[*a*]pyrene levels ranged from 0.09 to 2.3 ng/m^3. The highest levels of semivolatile aliphatic hydrocarbons and lead (Pb) confirmed the importance of mobile sources in this urban area during the day (Colombo et al. 1999; Bilos et al. 2001).

Other data reported for Argentinean air come from the Mendoza Province, a semiarid zone with an average humidity of about 35% and little rainfall (annual average, ~300 mm). Klaus et al. (1997) performed a study in two localities, in northern and southern Mendoza, measuring PAHs in PM10 particles. The concentrations varied from 0.592 to 4,879 ng/m^3 for phenanthrene, 0.03 to 0.16 ng/m^3 for anthracene, 0.128 to 0.778 ng/m^3 for fluoranthene, 0.19 to 0.46 ng/m^3 for pyrene, and from 0.0241 to 0.237 ng/m^3 for benzo[*a*]pyrene.

Chile is one of the most urbanized countries of South America, with almost 90% of its population living in urban areas (Tsapakis et al. 2002). Santiago, with approximately 5.8 million inhabitants, is a valley surrounded by mountains and has a ventilation level that inhibits the natural dispersion of polluting agents in air, resulting in high air pollution levels (Adonis and Gil 2000). This city is located in a zone of atmospheric stability characterized by low incidence of winds and annual precipitation less than 300 mm. All these factors combined with a high particle content of dust in suspension and smog has resulted in the formation of a thermal inversion layer in the atmosphere between 600 and 900 m over the city in the winter, reducing even more the dispersion of contaminates. All these factors have produced a large public health problem in Santiago.

In Santiago, the mobile sources of polluting agents include 9,000 buses and 37,000 diesel trucks using low-quality fuel with a high sulfide content similar to the fuel in developed countries. These factors are responsible for the high levels of respirable particle material PM10, carbon monoxide, and ozone, polluting agents that surpass the corresponding air quality standards at many times during the year. Sienra et al. (2002) found higher PAH concentrations in Santiago in the winter, where the highest proportions correspond to benzo[*e*]pyrene and benzo[*a*]pyrene with average values of 5.28 ng/m^3 and 6.37 ng/m^3, respectively.

Tsapakis et al. (2002) compared the PM2.5 fraction between the city of Temuco, a city in southern Chile with fewer than 300,000 inhabitants, and the greater Santiago area found significant differences between both localities. In Temuco, the high molecular weight PAHs such as benzo[*a*]-anthracene (BaA), chrysene (Chry), pyrene, benzofluoranthene (BFluo), and benzo[*a*]pyrene (BaPy), potential carcinogens, are the most abundant species of the aromatic fraction, surpassing to a great extent the concentrations observed in Santiago (Fig. 2). The major sources of PAHs in Temuco are household wood combustion stoves, whereas the sources in Santiago are related to vehicle sources. Temuco displays greater firewood consumption for domestic heating in comparison with Santiago. Authorities have sought to promote the replacement of wood and oil by gas and have introduced catalytic converters for unleaded gas cars.

PAHs were measured in air samples from Santiago by Adonis and Gil (2000). These data represent the only published information on the PAH

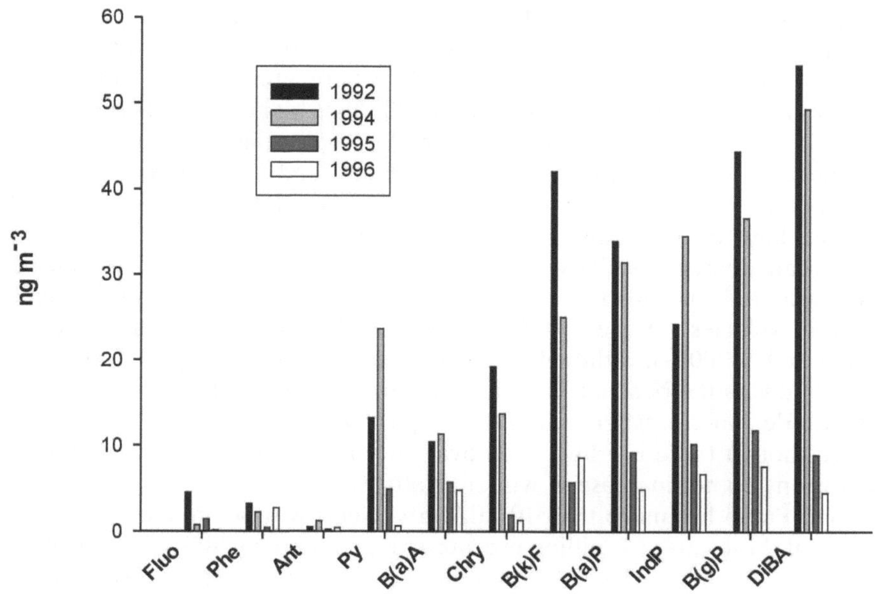

Fig. 2. Temporal trend of polycyclic aromatic hydrocarbons (PAHs) in PM10 in Santiago (Chile). Average concentrations are expressed as ng/m^3. See text for PAH abbreviations.

temporal trend in air particulate matter (PM10). Reported values are higher than values for other urban areas in the region, with concentrations ranging from 43 to 294 ng/m^3. The same authors have identified a clear seasonal PAH air pollution pattern with higher concentrations in winter and lower ones in the summer (Adonis et al. 2003).

Higher values were detected in 1992 (Gil et al. 2000), with a net decline of PAHs in recent years (see Fig. 2). Sienra et al. (2002) found that the total concentrations of PAHs in PM10 were of the order of 14.2 ng/m^3, a lower value than 729 ng/m^3 measured in 1991 by Didyk et al. (2000). Benzo[*a*]pyrene levels averaged 9.2 and 4.9 ng/m^3 for 1995 and 1996, respectively. Other carcinogenic PAHs detected included benzo[*k*]fluoranthene (23.5–4.3 ng/m^3 for 1992) and dibenzo[*a,h*]anthracene (54.5, 49.4, and 4.5 ng/m^3 for 1992, 1994, and 1996, respectively). The authors suggest that except for the reduction of air PAH levels, values are still high, presenting a potential human health risk.

The high PAH levels in winter, compared with those for spring and summer, are explained by the increased use of woodstoves, vehicular traffic, and meteorological conditions, which are less favorable for the dispersion of the polluting agents. On the other hand, during the summer the physicochemical and meteorological factors can significantly affect the atmospheric degradation of some PAHs, contributing to lower levels.

Kavouras et al. (1999) analyzed the PAH concentrations in the PM2.5 fraction, finding very different results to those reported by Didyk et al. (2000) for PM10 during winter periods, where great differences in the benzo[*a*]pyrene concentrations, close to 40 times higher, were observed. In Brazil, the principal urban PAH sources are vehicular emissions, domestic heating, waste combustion, oil refinement, and aluminum production (Fernandes et al. 2002). High urban PAH levels in Brazil can be explained by the high traffic levels, transport from industrialized areas located in suburban areas, and a high level of total suspended particles. Rio de Janeiro is considered one of the 20 largest cities in the world, with more than 1 million vehicles. In the last few years, gasoline sales have dramatically increased (~200%), although they contain a certain amount of ethanol, which lowers the benzene emissions. Niteroi is a tropical city located next to Rio de Janeiro, where summer is typically rainy and the winter is dry. Even though there is a lower removal rate of air pollutants, the pollutant emissions do not increase in winter (Netto et al. 2002).

Sao Paulo has more than 10 million vehicles with 4.7 million domestic cars. Of these cars, 3.7 million use a mixture containing 78%–80% gasoline and 22% ethanol, and 300,000 camions use diesel. Because of these characteristics and thermal inversion, PAHs are trapped within these strata.

Vasconcellos et al. (2003) performed a very complete study in the Sao Paulo area in three localities, Cidade Universitaria (urban traffic, gas, ethanol, diesel use), Cotia (diesel, truck traffic), and Agua Funda (refinery), where the last is an area with much vegetation, located at 20 km from an industrial park including a refinery. These authors found a different PAH pattern in air depending on the combustion source used: pyrene, chrysene, and fluoranthene are emitted by vehicles using gasoline, while chrysene, benzo[*a*]anthracene, and pyrene are emitted by vehicles consuming gasoline and diesel.

PAHs have also been sought in air samples from Brazil. Air concentration and total (wet and dry) deposits were determined in Salvador (Bahia State), Amazon State, Araraquara (São Paulo State), São Paulo city, Cubatão (São Paulo State), Londrina, Niteroi, and Candioca (Río Grande do Sul and Porto Alegre). Higher values were detected in the industrial area of Cubatão (55 ng/m^3), in the city of São Paulo where the concentrations ranged from 3 to 15 ng/m^3 compared with 0.003–1.5 ng/m^3 in other regions (Vasconcellos 1996; Vasconcellos et al. 1998; Beretta 2000; Franco 2001).

In Brazil, PAHs in the atmosphere of Araraquara (an agricultural area heavily devoted to sugar cane plantation and combustion) indicated unexpected low values for these compounds. This behavior was explained assuming that most of the fires in these plantations occurred in a smoldering fashion (low temperature), which is not the most favorable condition for PAH formation. Additionally, the PAHs were not separated from other pyrogenic sources, making data interpretation difficult. The main PAHs found were benzo[*b*]fluoranthene and benzo[*k*]fluorantene (Franco 2001).

Results presented in Table 1 are gaseous-phase values for winter 1994 and summer 1995, when 25 samples were analyzed for total PAHs. The results for benzo[*a*]anthracene ranged from 15 to 732 pg/m^3. The occurrence of substantial levels of certain PAH congeners and methyl-PAH derivates in airborne particles collected in the Amazonian forest in August and September 1993 is suggestive of emissions from extensive forest fires in that area. Indeed, a similar PAH pattern was detected on particles emitted by biomass combustion carried out under field and controlled conditions. The PAH distribution recorded in the rainforest was rather different from that observed in urban (São Paulo State, Brazil, and Rome, Italy) and suburban (Montelibretti, Italy) samples, because the airborne particulates came from both forest combustion and motor vehicle emission. Total PAH levels in the Amazonian forest were surprisingly high when compared with those commonly found in suburban, agricultural, and forest areas of Europe and North America.

A study performed in a rural community in winter 1991 in southern Brazil investigated the impact of wood-burning stoves on indoor air quality. The PAH, NO_2, and suspended particulate matter (SPM) concentrations were monitored in houses using woodstoves, and the results were compared with concentrations obtained in houses equipped with gas stoves. As expected, a higher ($P < 0.01$) concentration of PAHs and much higher ($P = 0.07$) concentrations of SPM existed when woodstoves were used. In

Table 1. Average Concentrations (ng/m^3) in PM10 Air Saimples Collected in Urban Areas of Brazil.[a]

Fluo	Phe	Ant	Py	BaA	Chry	Bfluo	B(e)P	B(a)P	IndP	BgP	DiBA	BPy	Sampling site/period
ND	8	ND	ND	12.7	4.3	7	12.8	2.8	0.8	3.2	3	0.2	Sao Paulo/ summer[b]
0.07	0.02	0.02	0.08	0.40	0.40	0.40	0.33	0.28	0.49	0.43	0.06	ND	Sao Paulo[c]
0.14	0.47		0.2	0.59	43	2.55	0.6	1.32	4.71	1.73	ND	ND	Sao Paulo[d]
0.49	0,32	0.03	0.54	0.3	0.59	1.11	0.49	0.4	0.76	ND	0.07	ND	Niteroi/winter[e]
0.3	0,27	0.03	0.8	0.13	0.24	1.55	0.34	0.33	0.86	ND	0.06	0.12	Niteroi/summer[e]
0.283	1.791	0.250	0.478	0.360	0.355	0.218	ND	0.417	0.296	0.291	0.051	ND	Mandioca[f]
0.68	0.68	0.131	0.324	0.753	0.525	1.236	ND	0.661	0.815	1.38	0.335	ND	Porto Alegre[g]

ND, below the detection limit.
[a]See text for polycylic aromatic hydrocarbon (PAH) abbreviations.
[b]De Martinis et al. 2002.
[c]Vasconcellos et al. 2003.
[d]Bourotte et al. 2005.
[e]Netto et al. 2002.
[f]Dallarosa et al. 2005a.
[g]Dallarosa et al. 2005b.

Table 2. Average Concentrations of Selected PAHs ($\mu g\,kg^{-1}$ d.w.) in Soils from Different Ecological Zones of Brazil and Chile.

Naph	Acen	Fluo	Phen	Ant	Pyr	BaA	Chry	BaP	IndP	DiBA	BgP	Ecological zone
53.67	0.23	2.70	20.00	0.31	4.33	0.12	0.79	0.19	0.12	0.11	0.18	Amazon (Br)
34	0.82	1.8	13	0.56	2.9	0.37	2.1	0.17	0.6	0.12	0.43	Pantanal (Br)
35	0.405	0.665	6	0.29	1.67	0.325	1.615	0.41	0.43	0.135	0.37	Cerrado (Br)
13	1.1	2.3	41	1	9.2	1.2	4.2	1.4	2.9	0.24	3.7	Rainforest (Br)
3.7	0.26	0.92	11	0.21	1.3	0.13	0.61	0.15	0.35	0.02	0.47	Savanna (Br)
6.3	ND	0.9	3.4	2.4	1.2	0.3	1.9	0.6	0.2	0.2	0.9	Natural coastal (Ch)
4.3	ND	1	2	0.5	3.9	1.1	2.3	2,2	1.6	0.4	1.9	Forestry (Ch)
17.5	ND	1	47.8	4	39.6	4.1	6.8	8,5	6.1	0.4	12.7	Industrial (Ch)
3.5	ND	1,1	2.4	0.4	3.3	1	2.4	2,9	1.8	0.5	2.3	Natural mountain (Ch)

Source: Wilcke et al. 2003; Barra et al. 2005a.

contrast, NO_2 concentrations were slightly higher in houses with gas stoves. These parameters were minimally affected by smoking, outdoor air pollution, or other emissions from indoor combustion products. Results appear to support the hypothesis that domestic wood-burning stoves are risk factors for some upper digestive and respiratory tract cancers in Brazil (Hamada et al. 1992).

A comparison of air PAH levels in urban areas of Argentina, Brazil, and Chile (Fig. 3) indicates that reported levels are by far higher in Santiago, although fluoranthene and benzofluoranthene are higher in urban areas of Brazil. Most of the Chilean data come from Santiago, which is surrounded by mountains that prevent good air circulation and there is a huge concentration of population. Most of the detected PAHs in particulate matter in Santiago are derived from motor vehicles. Meanwhile in Brazil, combustion of a gasoline–ethanol mixture (20%–25%) in vehicles releases less concentrations of PAHs into the atmospheric environment. Additionally, there is less wood consumption for heating in the tropical countries.

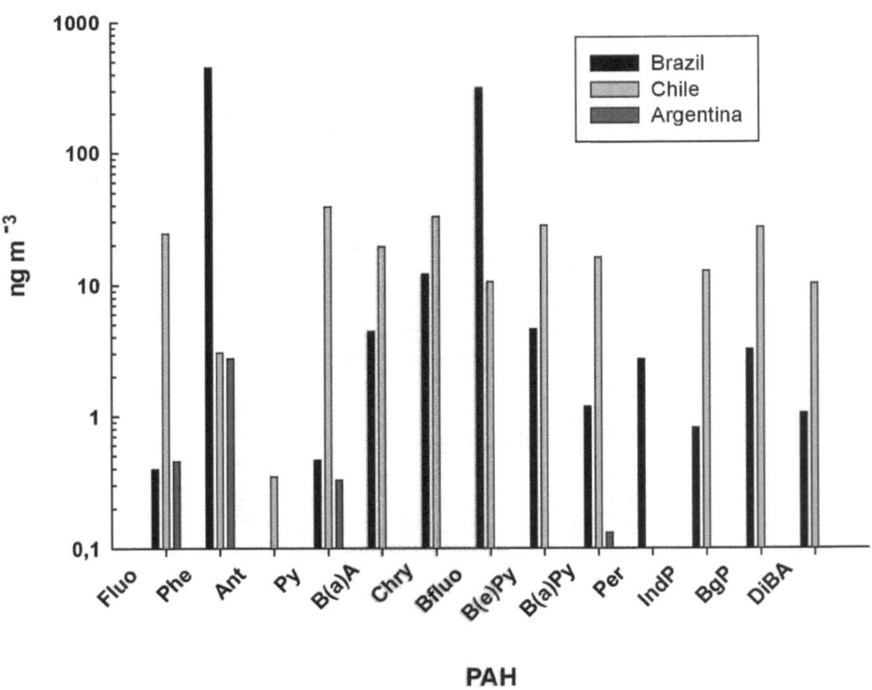

Fig. 3. Average PAH concentrations in air particulate samples (PM10) for several PAHs in Argentina, Brazil, and Chile.

B. Soil

Knowledge about PAH sources helps us understand the global dynamics of their movement through different environmental compartments. Soils could be acting as a sink for many PAHs within their global circulation. Most of the PAHs are released in temperate areas as fossil fuel combustion by-products, but other processes include diagenesis, forest fires, and volcanic activities (Sims and Overcash 1983). Reports have shown that biological processes release low molecular weight PAHs in tropical environments, contributing to their global discharge (Wilcke 2000).

PAH concentrations in urban soils come principally from industrial and vehicular sources, and most of the PAHs sources in tropical environments are not related to fossil fuel combustion processes (Wilcke 2000); consequently, forest fires or biological processes could be important sources of PAHs in soils in Brazil. Wilcke (2000) reported PAH concentrations from different ecological zones in Brazil, and their major findings relate to an abundant pattern of individual PAH compounds, where the most common are naphtalene, phenanthrene, or perylene. In temperate soils, the most abundant as fossil fuel tracers are chrysene, fluoranthene, and pyrene (Wilcke et al. 2003), which are minor contributors to the PAH burden in tropical soils.

The detection of high levels of naphthalene in termite nests (Wilcke 2000) opens the possibility that some sources could be related to termite feeding or the production of such compounds in the nest. In tropical mineral soils, concentrations in the first centimeter are in the range of 42–65 $\mu g\,kg^{-1}$, lower than in the topsoil samples, which ranged between 6.5 and 397 $\mu g\,kg^{-1}$. These PAH concentrations are lower than those in temperate zone soils, where Wilcke (2000) reported average concentrations of the order of 284–4420 $\mu g\,kg^{-1}$ in soils from agricultural, forest, and urban areas.

Barra et al. (2005) reported, for the first time, PAH concentrations in Chilean soils from different ecological zones. When a comparison is made with Brazilian soils, an industrialized soil from Chile is well above levels detected in Brazil. For example, benzo[*a*]anthracene, chrysene, benzo[*a*]pyrene, benzo[*ghi*]perylene, and indene levels are higher in Chile. It is important to note that this area is located within a forest plantation and near a sawmill; the high concentration in this site is clearly related to anthropogenic activities, and PAH incorporation into the soil could be the result of forest fires and the use of wood residues as combustion sources.

C. Water

PAH concentrations in water are relatively high (average, $16 \pm 12\,\mu g/L$), although the database is not homogeneous. Lower concentrations (1.8–12 $\mu g/L$) have been determined by gas chromatography-flame ionization detection (GC-FID) in the Uruguay and Rio de la Plata rivers, whereas the highest levels (8–41 $\mu g/L$) correspond to fluorimetric determinations in coastal marine Patagonian waters impacted by crude oil extraction activities (Esteves and Commendatore 1993). The Canadian water quality guidelines for 11 individual PAHs range from 0.012 to 5.8 μg/L (mean, 1.7 ± 2.1 μg/L; sum, 18.2 μg/L).

In Brazil, some studies have been performed in the Paraiba do Sul River near Rio de Janeiro; the only PAH detected of six analyzed by gas chromatography-mass spectrometry (GC-MS) methods was benzo[*a*]pyrene in levels reaching 0.255 μg/L near a urban town. Brazilian benzo[*a*]pyrene limits are 0.01 μg/L in drinking water (Azevedo et al. 2004).

D. Sediment

Most of the PAH reports compiled correspond to harbors and ports in heavily impacted areas and thus present a huge variability (0.1–286,000 μg/kg). The largest mean PAH value is very high (29,520 ± 63,492 μg/kg), exceeding by several orders of magnitude the Canadian guidelines for individual PAHs (5.9–111 μg/kg for 13 PAHs; sum, 468 μg/kg). Most affected areas correspond to the intensive traffic in the Paraná and Río de la Plata rivers in Uruguay, especially close to heavily populated areas such as Buenos Aires and Montevideo, the Argentine Patagonian coastal area, where crude oil extraction and transport are very active, and in the Tiete River and surrounding environments close to Sao Paulo in Brazil. Also, Torres et al. (2002) sampled and analyzed bottom sediment samples from the Paraíba do Sul river watershed for PAHs. The contamination found is moderate and is related to steelworks (Torres et al. 2002). Solid waste from different industrial origins contains at least 1 of the 16 EPA PAHs. Muniz et al. (2004) analyzed sediment samples from the Montevideo harbor in Uruguay, reporting levels from 1.56 to 90.44 $\mu g\, g^{-1}$, which were attributed to petroleum and derivates. The highest levels were reported for phenanthrene, fluoranthene, and pyrene, respectively. According to the authors, a preliminary risk analysis indicates that the concentrations recorded present risk to benthic organisms. In another study, at the Bahia de Todos os Santos in Brazil, PAH levels in sediments varied from 8 to 4.163 $ng\, g^{-1}$ (Venturini and Tomasso 2004). The authors suggest trophic chain changes produced by exposure in both deposit feeders and carnivores.

Palma-Fleming et al. (2004) measured PAHs in sediment samples in the Valdivia estuary in southern Chile, and the levels reported corresponded to medium PAH pollution, ranging from 6.9 to 74.1 $ng\, g^{-1}$ (dry weight, d.w.), and also were highly variable during different seasons. The main PAH sources were of pyrolitic origin. The authors mention that in that area woodstove combustion could be the main source of the PAH detected, even though a harbor is located near the study area and the marine traffic is quite intensive.

Patagonian total fluorimetric PAHs concentrations (29,500 ± 59,300 $\mu g\, kg^{-1}$) are very high, considering that these are mostly coarse sediments

(0.1%–58% fines) with low organic contents (0.4%–3.6%). However, CG-FID evaluation of the aliphatic composition of these samples indicated a clear petrogenic signature in most contaminated sites from the San Jorge Gulf, which is severely affected by oil production and transport activities (Commendatore et al. 2000). PAH concentrations in the Río de la Plata estuary are generally lower but present some critical values in bays, harbors, and ports.

In Uruguay, Muñiz et al. (2004) reported that total PAH concentrations in sediments from the Montevideo harbor ranged from 1.56 to 90.44 $\mu g\,g^{-1}$, similar to those found in severely polluted areas such as Santos in Brazil (Nishigima et al. 2001). They performed two surveys in this harbor and concluded that the source is the combustion of oil and oil derivatives because they also found alkylated aromatic hydrocarbons in the samples they analyzed.

More recently, PAH levels in sediment cores from pristine Andean lakes have been reported (Quiroz et al. 2005; Barra et al. 2006), suggesting that the diagenetic origin of PAHs predominates in the composition of such sediments; this observation indicates that transport and deposition in such remote areas remains very limited in the past 50 years, where the levels observed were comparable to levels found in lake sediments from the Northern Hemisphere. For comparative purposes, such areas could represent the natural PAH composition in sediments.

E. Foodstuffs

The main environmental effect of PAHs is related to their health effects and especially their carcinogenic properties. Their toxic and carcinogenic potential and the high stability of most PAHs make them an especially interesting group when searching for contaminant presence in many foods. This contamination can result from sorption from a contaminated environment or from different food preparation methods, especially in noncontrolled and informal food processing, such as in street markets where wood or coal is burnt, a very common situation in several South American countries. After absorption in the body, they may be altered into substances that are able to damage the cell genetic material and initiate the development of cancer, although individual PAHs differ in their capacity to damage cells in this way.

According to Toledo and Camargo (1998), most corn oils of different brands that are produced and commercialized in Brazil may contain more than 1 ppm benzo[*a*]pyrene. Additionally, Noll and Toledo (1997), working with smoked food samples (meat), found that the benzo[*a*]pyrene levels are higher in homemade products than in commercial ones, with results that can reach 6.1 $mg\,kg^{-1}$.

Vegetable oils can be contaminated when the raw material is contaminated, such as when the plants receive airborne pollutants, are dried with smoke before extraction, or are contaminated by the solvents used in the extraction procedure (Pupin and Toledo 1996). Although this fact received special attention in the 1960s and 1970s in Europe and other developed countries, concern is now focused on the well-known carcinogen benzo[*a*]pyrene and on the notable amounts of light PAH (3–4 rings) that have been found in olive oils, a typical cold-pressed product. To avoid this contamination, several technological upgrades are available, such as active carbon filtration or steam distillation processes. According to Toledo and Camargo (1998), most corn oils from different brands produced and commercialized in Brazil may have more than 1 $mg\,kg^{-1}$ of benzo[*a*]pyrene.

It is only in the few countries that possess specific food residue legislation, such as Germany, where the total content of light PAHs (<3–4 rings) should not exceed 25 ppm, whereas the content of the heavier PAHs (>5 rings) must remain below 5 $ppm\,kg^{-1}$. In Brazil, according to Noll and Toledo (1995), the presence of PAHs in charcoal-broiled meat is sometimes above such limits, depending whether it was cooked with or without the fat, indicating the importance of controlling the cooking method to minimize carcinogenic compound formation. The distance from the heat source is another very important issue when studying such residues in relation to barbecues (Noll and Figueiredo 1997).

Sugarcane plantations are another well-known PAH source in Brazil, which results from the traditional cultural practice of burning fields for an easier harvest. This contamination could be detected in most of the yielded products, including the sugarcane spirit, the "cachaça" (Serra et al. 1995).

F. Biota

PAH data in biota are available for a limited number of samples, including the already-mentioned Mussel Watch Program. Data shown for PAHs (Fig. 4) are consistent with the pattern observed for other persistent pollutants (such as polychlormated biphenyls, PCBs) (Barra et al. 2005b). In less polluted areas, background levels in bivalves are below 10 $mg\,kg^{-1}$ lipids, ranging from 10 to 50 $mg\,kg^{-1}$ in moderately polluted samples, and above 200 mg kg^{-1} in some impacted sites such as Punta Arenas (Chile), Recife (Brazil), Concepción (Chile), and Bahía Camarones, Río de la Plata, and Bahía Blanca (Argentina). Elevated concentrations in some of these sites can be associated with offshore oil production and petrochemical activities (Argentine Patagonia and Brazil). A similar pattern was observed in the Santos coastal area (São Paulo state) where bivalves collected at 26 points in an area under influence of a petrochemical complex presented values as high as 860 $ng\,g^{-1}$ (mainly naphthalene) according to the CETESB report (UNEP, 2002).

Mussels were also analyzed by Palma-Fleming et al. (2004) in Corral Bay in southern Chile, finding an intermediate level of pollution with a clear pyrolitic signal. The levels found were also highly variable within seasons

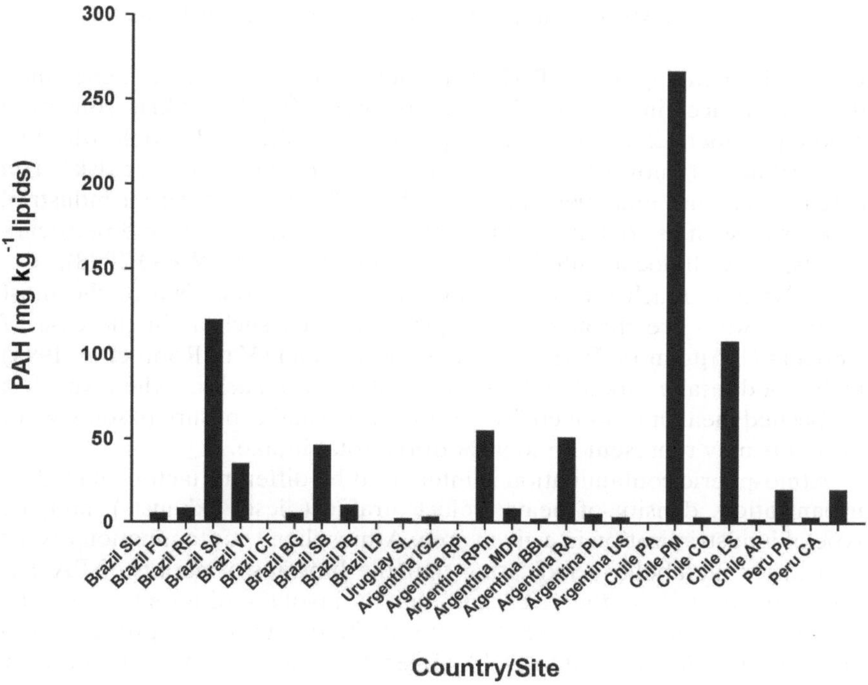

Fig. 4. PAH data from the Mussel Watch Program in South America (data from Farrington and Tripp 1995).

and years, ranging from 3.1 to 390 ng g^{-1} d.w., indicating a preferential accumulation of low molecular weight PAHs with levels quite similar to those found in sediments.

Toro et al. (2004) analyzed PAH levels in the Chilean giant mussel (*Choromitylus chorus*) in three areas with different anthropogenic impact. They found significant differences in concentration ranges, where the polluted San Vicente Bay presented differences of two to three orders of magnitude in comparison with the levels found in pristine marine systems from Southern Chile (3880 ng g^{-1} vs. 82 ng g^{-1} d.w., respectively). The patterns found also indicate a clear petrogenic origin.

Other approaches for investigating the PAH levels in freshwater and marine fish that have been found to be very cost-effective consist of determining PAH metabolites in fish bile as a biomarker of PAH exposure with fluorescence spectrometry (Barra et al. 2001; Fuentes-Rios et al. 2005; Inzunza et al. 2006). This method presents itself as a valuable alternative for analyzing fish exposure to PAH-like compounds in countries with limited resources.

G. Industrialization, Urbanization, and Health Issues

Generally speaking, some PAH compounds are related to a higher incidence of cancer in humans. Because of their ubiquity, PAHs constitute a potential menace to the entire population's health, and people who live or work near a known PAH source are exposed to even higher risks. The carcinogenic and mutagenic effects of the effluents of different industrial processes, such as coal and aluminum industries, refineries, and petrochemicals, are well documented (Van Schooten et al. 1995; WHO 1998).

PAHs may reach the human body in several ways. Two of the most common ways are through the respiratory tract, such as in the case of smokers (1–5 μg more PAHs/d than a nonsmoker) (Van Rooij et al. 1994) or by food intake (about 96% of the total intake), such as when we eat a barbecued meal or a smoked fish. However, dermal exposure in some work incidents may represent up to 90% of the total intake.

Atmospheric contamination is influenced by different factors, including urbanization, density of heavy vehicle traffic (diesel exhausts), and the type of industrialization in a given area. Atmospheric contamination results in a faster metabolism, indicating that the biotransformation of PAH is not only located in the liver and that it involves different enzymatic pathways (Ishidate et al. 1980). Some of these routes may produce diol-epoxides, which may bind to DNA molecules, causing transcription errors. This phenomenon is known as activation and is related to the mutagenic properties of some of the 4- or 5-ring PAHs. Bioaccumulation is less prone to occur.

Cubatao is one of the most industrialized towns of Brazil. In this city, a study performed to analyze suspicious birth defects did not find significant differences between Cubatao and other towns with respect to observed cases of malformations. However, the suspicions about the effects of contamination and other problems related to benzene and organochlorine exposure helped to start a very important project to clean up the Cubatao environment (Augusto and Novaes 1998). Additionally, Navy workers in Brazil may have higher mortality due to certain cancer types when compared to other population groups. The authors of the study suggest exposure to a variety of chemicals that does not includes PAHs (Silva et al. 2000).

Aluminum production and anode baking is currently the largest single emission source for individual PAHs, with these activities being concentrated near the ore sources in Brazil (Bauxite: Al Mining and smelting at Para, Pernambuco, and Maranhao states). Other sources are related to fossil fuel burning, road traffic, and domestic wood combustion, where wood combustion is of particular importance in the Brazilian countryside. Wood treatment with creosote is considered a significant source of the lighter PAHs. Bitumen is a probable source of benzo[*a*]pyrene and other PAHs.

The increase in urbanization in many South American countries will lead to higher PAH pollution levels, and the health implications have received little attention by the regulatory authorities and decision makers.

Summary

Pollution of the environment with polycyclic aromatic hydrocarbons (PAHs) should be a global concern, especially in urbanized areas. In South American countries, where notable increase in urban populations has been observed in the past few years, reliable information about the pollution status of these urban environments is not always easily accessible, and therefore an effort to collect updated information is required. This review attempts to contribute by analyzing the existing information regarding environmental levels of PAHs in some South American countries.

A regional trend for environmental PAH information is an uneven contribution, because some countries, such as Bolivia, Peru, Paraguay, and Ecuador, have reported no information at all in the scientific literature, reflecting to a certain extent the different patterns of economic, technical, and scientific development.

PAH air monitoring is one of the areas that has received the most attention during the last few years, mainly in Brazil, Chile, and Argentina, where data represent a few geographical areas within the region. PAH levels in air from some urban areas in Argentina, Brazil, and Chile, considered moderate to high (100–1000 ng/m^3), are probably among the highest values reported in the open literature. Urbanization, vehicle pollution, and wood fires are the principal contributors to the high reported levels. In more temperate areas, a clear distinction is observed between summer and winter levels.

PAH monitoring in soils is very limited within the region, with few data available, and most information indicates widespread pollution. In Brazil, values for many representative ecosystems were found. In Chile, data from forestry and agricultural areas indicate in general low concentrations, in spite of a relatively high detection frequency. Pollution levels in soils are highly dependent on their closeness to PAH sources and certain cultural practices (agricultural burnings, forest fires, etc.).

Water PAH levels are rarely reported in the scientific literature for South American countries. Few data were available, even though many regulatory agencies perform routine analysis of hydrocarbons in waters. No information was found specifically related to PAH compounds, which could indicate generally low PAH levels in waters.

Regional PAH information for sediments also indicates higher levels. Overall, as observed for water, sediment data indicate a complex situation in densely populated areas affected by urban–industrial inputs where high PAH levels are found. In contrast, in remote areas a typical profile of diagenetic PAHs dominates. Concentrations are greatly variable and are principally related to several highly contaminated sites in Argentina and Brazil (hot spots) with levels four to five orders of magnitude higher.

Even though PAHs have carcinogenic properties, little attention has been paid to the analysis of aquatic organisms except in the case of bivalves. As observed for other environmental receptors, the regional data distribution is uneven and is heavily centered in coastal environments and in a few countries (Argentina, Brazil, Chile, and Peru). The most comprehensive PAH monitoring program in the South American coastal environment is the Mussel Watch. Baseline PAH concentrations range from 200 to 700 $\mu g\,kg^{-1}$ lipids in unpolluted sites; from 1,000 to 3,000 $\mu g\,kg^{-1}$ in moderately contaminated sites; and from 4,000 to 13,000 µg/kg lipids in the most affected bivalves that come from areas of Río de la Plata (Argentine side), Recife (Brazil), and Punta Arenas (Chile).

Critical data gaps exist with respect to PAH analysis in biota, including humans, in foodstuffs, and subsequent effects. Considering the high levels reported in the air compartment, risk assessment procedures in highly populated areas need to be performed. Additionally, few countries within the region have information on PAH levels. In these countries, this type of analysis needs to be performed, and the laboratory capacity needs to be built to assure the accomplishment of these objectives.

Acknowledgments

This work was partially supported by FONDECYT Grant No. 1050647 (Chile) awarded to Ricardo Barra. The RED PROSUL CNPq, Latin American Network for Persistent Toxic Substances Studies, Contract Number 490328/2004-0 (Joao Machado Torres) UFRJ, Conicyt-CNPq project N° 2005-076, is also gratefully acknowledged.

References

Adonis M, Gil L (2000) Polycyclic aromatic hydrocarbon levels and mutagenicity of inhalable particulate matter in Santiago, Chile. Inhal Toxicol 12: 1173–1183.

Adonis M, Martínez V, Riquelme R, Ancic P, González G, Tapia R, Castro M, Lucas D, Berthou F, Gil F (2003) Susceptibility and exposure biomarkers in people exposed to PAHs from diesel exhaust. Toxicol Lett 144:3–15.

Ares J, Zavatti J (1993) Comparative analysis of emissions and diffusion of air PAHs at a coastal arid site (Patagonia, Argentina). Bull Environ Contam Toxicol 50: 333–339.

Augusto LGS, Novaes TCP (1998) Ação médico-social no caso do benzenismo em Cubatão, São Paulo: uma abordagem interdisciplinar. Cad Saúde Pública 15(4):729–738.

Azevedo D, Gerchon E, dos Reis E (2004) Monitoring of pesticides and polycyclic aromatic hydrocarbons in water from Paraiba Do Sul river. J Brazil Chem Soc (15)2:292–299.

Barra R, Sanchez-Hernandez JC, Orrego LR, Parra O, Gavilan JF (2001) Bioavailability of PAHs in the Biobio River (Chile): MFO activity and biliary fluorescence in juvenile *Onchorhynchus mykiss* 45:439–444.

Barra R, Popp P, Quiroz R, Bauer C, Cid H, Tumpling W (2005a) Persistent toxic substances in soils and waters along an altitudinal gradient in the Laja River Basin, Central Southern Chile. Chemosphere 58:905–915.

Barra R, Colombo JC, Eguren G, Gamboa N, Jardim WF, Mendoza G (2005b) Persistent organic pollutants (POPs) in eastern and western South American countries. Rev Environ Contam Toxicol 185:1–33.

Barra R, Popp P, Quiroz R, Treutler HC, Araneda A, Bauerb C, Urrutia R (2006) Polycyclic aromatic hydrocarbons fluxes during the past 50 years observed in dated sediment cores from Andean mountain lakes in central south Chile. Ecotoxicol Environ Saf 63:52–60.

Beretta M (2000) Adequação de protocolo analítico para hidrocarbonetos petrogênicos na atmosfera do recôncavo baiano. Tese de doutorado. Química Analítica, Instituto de Química, UFBA, Salvador, pp 139–140.

Bilos C, Colombo JC, Skorupka C, Rodriguez Presa MJ (2001) Sources, distribution and variability of airborne trace metals in La Plata City area, Argentina. Environ Pollut 111:149–158.

Bourotte C, Forti MC, Taniguchi S, Bícego MC, Lotufo PA (2005) A wintertime study of PAHs in fine and coarse aerosols in São Paulo city, Brazil. Atmos Environ 39(21):3799–3811.

Catoggio JA, Succar SD, Roca AE (1989) Polynuclear aromatic hydrocarbons content of particulate matter suspended in the atmosphere of La Plata, Argentina. Sci Total Environ 79:43–58.

Colombo JC, Landoni P, Bilos C (1999) Sources, distribution and variability of airborne particles and hydrocarbons in La Plata area, Argentina. Environ Pollut 104:305–314.

Commendatore MG, Estevez JL, Colombo JC (2000) Hydrocarbon in coastal sediments of Patagonia, Argentina: levels and probable sources. Mar Pollut Bull 40(11):989–998.

Dallarosa JB, Teixeira EC, Pires M, Fachel J (2005a). Study of the profile of polycyclic hydrocarbons in atmospheric particles (PM_{10}) using multivariate methods. Atmos Environ 39:6587–6596.

Dallarosa JB, Monego JG, Teixeira EC, Stefens JL, Wiegand F (2005b) Polycyclic aromatic hydrocarbons in atmospheric particles in the metropolitan area of Porto Alegre, Brazil. Atmos Environ 39:1609–1625.

De Martinis BS, Okamoto RA, Kado NY, Gundel LA, Carvalho LRF (2002) Polycyclic aromatic hydrocarbons in a bioassay-fractionated extract of PM 10 collected in Sao Paulo, Brazil. Atmos Environ 36:307–314.

Didyk BM, Simoneit BRT, Pezoa LA, Riverso ML, Flores AA (2000) Urban aerosol particles o Santiago, Chile: organic content and molecular characterization. Atmos Environ 34:1167–1179.

Esteves JL, Commendatore M (1993) Total aromatic hydrocarbons in water and sediments in coastal zone of Patagonia, Argentina. Mar Pollut Bull 26(6):341–342.

Farrington JW, Tripp BW (1995) International Mussel Watch Program. Initial implementation phase. Final report. NOAA Tech Memo NOS ORCA 95.

Franco A (2001) Determinação de hidrocarbonetos poucos voláteis no material particulado da atmosfera de Araraquara. Instituto de Química. UNESP, Araraquara.

Fernandes M, Brickus L, Moreira JC, Cardoso JN (2002) Atmospheric BTX and polyaromatic hydrocarbons in Rio de Janeiro, Brazil. Chemosphere 47:417–425.

Fuentes-Rios D, Orrego R, Rudolph A, Mendoza G, Gavilán JF, Barra R (2005) EROD activity and biliary fluorescence in *Schroederichthys chilensis* Guichenot 1848): biomarkers of PAH exposure in coastal environments of the South Pacific Ocean. Chemosphere 61:192–199.

Gil L, King L, Adonis M (2000) Trends of polycyclic aromatic hydrocarbon levels and mutagenicity in Santiago's inhalable airborne particles in the period 1992–1996. Inhal Toxicol 12:1185–1204.

Hamada GS, Kowalski LP, Murata Y, Matsushita H, Matshuki H (1992) Wood stove effects on indoor air quality in Brazilian homes: carcinogens, suspended particulate matter, and nitrogen dioxide analysis. Tokai J Exp Clin Med 17: 145–153.

Hafner W, Carlson D, Hites R (2005) Influence of local human population on atmospheric polycyclic aromatic hydrocarbon concentrations. Environ Sci Technol 39(19):7374–7379.

Inzunza B, Orrego R, Peñalosa M, Gavilan JF, Barra R (2006) Analysis of CYP4501A1, PAHs metabolites in bile, and genotoxic damage in *Oncorhynchus mykiss* exposed to Biobio river sediments, Central Chile. Ecotoxicol Environ Saf 62:52–60.

Ishidate K, Tsuruoka M, Nakasawa Y (1980) Induction of choline kinase by polycyclic aromatic hydrocarbons. Biochem Biophys Res Comm 96(2):946–952.

Kavouras IGJ, Koutrakis LP, Stephanou EG, Oyola P (1999) Measurement of particulate aliphatic and polynuclear aromatic hydrocarbons in Santiago de Chile: source reconciliation and evaluation of sampling artifacts. Atmos Environ 33:4977–4986.

Klaus-Dieter W, WeiBflog L, Paladini E, Gantuz M, Guerreiro P, Puliafito C, Schurmann G (1997) Emmission patterns of airborne pollutants in Argentina and Germany. II. Biomonitoring of organochlorine compounds and polycyclic aromatics. Chemosphere 34(12):2505–2518.

Muniz P, Danulat E, Yannicelli B, Garcia-Alonso J, Medina G, Bicego MC (2004) Assessment of contamination by heavy metals and petroleum hydrocarbons in sediments of Montevideo Harbour (Uruguay). Environ Int 29(8):1019–1028.

Netto Pereira AD, Barreto RP, Moreira JC, Arbilla G (2002) Polycyclic aromatic hydrocarbons in total suspended particulate of Niteroi, RJ, Brazil: a comparison of summer and winter samples. Bull Environ Contam Toxicol 69:173–180.

Nishigima FN, Webor RR, Bicego MC (2001) Aliphatic and aromatic hydrocarbons of Santos and Cananeia SP, Brazil Mar Poll Bull 42(11):1064–1072.

Noll IB, Figueiredo MC (1997) Determinacion de benzo[a]pireno em asados tipo churrasco en Brazil. Inform Tecnol 8(2):177–181.

Noll IB, Toledo MCF (1995) Benzo[*a*]pireno en carnes defumadas pelos processos caseiro e industrial. Rev Esp Cienc Tecnol Aliment 35(2):209–215.

Noll IB, Toledo MCF (1997) Note. Polycyclic aromatic hydrocarbons in charcoal-broiled meat in Brazil. Rev Bras Toxicol 10(1):19–23.

Palma-Fleming H, Asencio AJ, Gutierrez PE (2004) Polycyclic aromatic hydrocarbons in sediments and mussels of Corral Bay, South Central Chile. J Environ Monit 6:229–233.

Pupin MA, Toledo MCF (1996) Benzo[*a*]pireno in olive oils on the Brazilian market. Food Chem 55(2):185–188.

Quiroz R, Popp P, Urrutia R, Bauer C, Araneda A, Treutler HC, Barra R (2005) PAH fluxes in the Laja Lake of south central Chile Andes over the last 50 years: evidence from a dated sediment core. Sci Total Environ 349(1–3):150–160.

Serra GE, Pupin MA, Toledo MCF (1995) Ensaios preliminares sobre a contaminação da cana de açúcar e derivados por hidrocarbonetos policíclicos aromáticos. Bol SBCTA 29(2):134–137.

Sienra MR, Rosazza NG, Préndez M (2002) Polycyclic aromatic hydrocarbons and their molecular diagnostic ratios in urban atmospheric respirable particulate matter. Atmos Res 75:267–281.

Silva M, Santana VS, Loomis D (2000) Mortalidade por câncer em militares da Marinha do Brasil. Rev Saúde Pública 34(4):373–379.

Sims RC, Overcash MR (1983) Fate of polynuclear aromatic compounds (PANs) in soil–plant systems. Residue Rev 88:1–68.

Toledo MCF, Camargo MSFO (1998) Benzo[*a*]pireno em óleos de milho produzidos e comercializados no Brasil. Ciênc Tecnol Aliment Int 18(1):73–76.

Toro B, Navarro JM, Palma-Fleming H (2003) Use of clearance rate in *Choromytilus chorus* (Bivalvia: Mytilidae) as a nondestructive biomarker of aquatic pollution. Rev Chil Hist Nat 76:267–274.

Toro B, Palma-Fleming H, Navarro JM (2004) Organic pollutant burden of the giant mussels *Choromytilus chorus* from the south-central Chilean coast. Chemosphere 55:267–275.

Torres JPM, Malm O, Vieira EDR, Japenga J, Koopmans GF (2002) Organic micropollutants on river sediments from Rio de Janeiro state, Southeast Brazil. Cad Saúde Pública Rio de Janeiro 18(2):277–488.

Tsapakis ME, Lagoudaki EG, Stephanou I, Kavouras G, Koutrakis P, Oyola P, Von Baer D (2002) The composition and sources of pm 2.5 organic aerosol in two urban areas of Chile. Atmos Environ 36:3851–3863.

UNEP (2002) Eastern and Western South America: Regional Report. United Nations Environment Programme Chemicals. UNEP, Geneva, Switzerland.

Van Rooij JG, Veeger MM, Bodelier-Bade MM, Scheepers PT, Jongeneelen FJ (1994) Smoking and dietary intake of polycyclic aromatic hydrocarbons as sources of inter-individual variability in the baseline excretion of 1-hydroxypyrene in urine. Int Arch Occup Environ Health 66(1):55–65.

Van Schooten FJ, Jongeneelen FJ, Hillebrand MJ, van Leeuwen FE, de Looff AJ, Dijkmans AP, van Rooij JG, den Engelse L, Kriek E (1995) Polycyclic aromatic hydrocarbon-DNA adducts in white blood cell DNA and 1-hydroxypyrene in the urine from aluminum workers: relation with job category and synergistic effect of smoking. Cancer Epidemiol Biomark Prev 4(1):69–77.

Vasconcellos PC (1996) Um estudo dobre a caracterização de hidrocarbonetos policíclicos aromáticos e seus derivados, e hidrocarbonetos alifáticos saturados em material particulado atmosférico proveniente de sítios urbanos, suburbanos e florestais. Tese de doutorado. Instituto de Química, Universidade Estadual de São Paulo.

Vasconcellos PC, Ciccioli P, Artaxo P, Brancaleoni E, Cecinato A, Frattoni M (1998) Determinacao dos hidrocarbonetos saturados e polic!ıclicos aromaticos na atmosfera Amazonica. Quım Nova 21:386–393.

Vasconcellos PC, Zacarias D, Pires MAF, Pool CS, Carvalho LRF (2003) Measurements of polycyclic aromatic hydrocarbons in airborne particles from the metropolitan area of São Paulo City, Brazil. Atmos Environ 37:3009–3018.

Venturini N, Tommasi LR (2004) Polycyclic aromatic hydrocarbons and changes in the trophic structure of polychaete assemblages in sediments of Todos os Santos Bay, Northeastern Brazil. Mar Pollut Bull 48:97–107.
WHO (1998) Environmental Health Criteria 202, selected non-heterocyclic polycyclic aromatic hydrocarbons, World Health Organization, Geneva, Switzerland.
Wilcke W (2000) Polycyclic aromatic hydrocarbons (PAHs) in soil: a review. J Plant Nutr Soil Sci 163:229–248.
Wilcke W, Amelung W, Krauss M, Martius C, Bandeira A, Garcia M (2003) Polycyclic aromatic hydrocarbons (PAH) patterns in climatically different ecological zones of Brazil. Org Geochem 34:1405–1417.

Manuscript received July 3, 2006; accepted July 6, 2006.

Rev Environ Contam Toxicol 191:23–45

Silver as a Disinfectant

Nadia Silvestry-Rodriguez, Enue E. Sicairos-Ruelas,
Charles P. Gerba, and Kelly R. Bright

Contents

I. Introduction

The antimicrobial effects of silver (Ag) have been recognized for thousands of years. In ancient times, it was used in water containers (Grier 1983) and to prevent putrefaction of liquids and foods. In ancient times in Mexico, water and milk were kept in silver containers (Davis and Etris 1997). Silver was also mentioned in the Roman pharmacopoeia of 69 B.C. (Davis and Etris 1997).

Communicated by Dr. Charles P. Gerba.

N. Silvestry-Rodriguez
Department of Agricultural and Biosystems Engineering, University of Arizona, Tucson, AZ 85721, U.S.A.

E.E. Sicairos-Ruelas, C.P. Gerba, K.R. Bright (✉)
Department of Soil, Water and Environmental Science, University of Arizona, Building 38, Room 429, Tucson, AZ 85721, U.S.A.

In 1884, silver nitrate drops were introduced as a prophylactic treatment for the eyes of newborns, and this became a common practice in many countries throughout the world (Wahlberg 1982) to prevent infections caused by *Neisseria gonorrhoeae* transmitted from infected mothers during childbirth (Klueh et al. 2000; Slawson et al. 1992). In 1928, the "Katadyn Process," based on the use of silver in water at low concentrations, was introduced (Krause 1928).

Silver ions have the highest level of antimicrobial activity of all the heavy metals. Gram-negative bacteria appear to be more sensitive than gram-positive species (Feng et al. 2000; Kawahara et al. 2000; Klueh et al. 2000). Kawahara et al. (2000) posited that some silver binds to the negatively charged peptidoglycan of the bacterial cell wall. Because gram-positive species have a thicker peptidoglycan layer than do gram-positive species, perhaps more of the silver is prevented from entering the cell.

Generally speaking, the observed bactericidal efficacy of silver and its associated ions is through the strong binding with disulfide (S–S) and sulfhydryl (–SH) groups found in the proteins of microbial cell walls. Through this binding event, normal metabolic processes are disrupted, leading to cell death. The antimicrobial metals silver (Ag), copper (Cu), and zinc (Zn) have thus found their way into a number of applications.

II. Applications and Uses

A. Drinking Water

Chlorine has been used as the principal disinfectant for drinking water since the early 1900s. In the 1970s, it was discovered that chlorination caused the formation of numerous chlorinated compounds in water, including trihalomethanes and other disinfection by-products (DPB), that are known to be hazardous to human health (Moudgal et al. 2000; Von Gunten et al. 2001). There is therefore a need to assess alternative disinfectants (Yahya et al. 1992).

Silver electrochemistry experiments suggest that silver may have potential as a chlorine alternative in drinking water disinfection in applications in which chlorine may be considered too hazardous (Pedahzur et al. 2000). Silver has been used as an effective water disinfectant for many decades (Kim et al. 2004), primarily in Europe (Russell and Hugo 1994). It has also been used to treat recycled water aboard the MIR space station and aboard NASA space shuttles (Butkus et al. 2004; Gupta et al. 1998).

Both the Environmental Protection Agency (EPA) and the World Health Organization (WHO) regard silver as safe for human consumption. Only argyria (irreversible skin discoloration) occurs with the ingestion of gram quantities of silver over several years or by the administration of high concentrations to ill individuals. There have been no reports of argyria or other toxic effects caused by silver in healthy persons (World Health

Organization 1996). Based on epidemiological and pharmacokinetic data, a lifetime limit of 10 grams of silver can be considered a No Observable Adverse Effect Level (NOAEL) for humans (World Health Organization 1996). In the United States, no primary standards exist for silver as a component in drinking water. The EPA recommends a secondary nonenforceable standard of 0.1 mg/L (100 ppb) (Environmental Protection Agency 2002). The World Health Organization (1996) has stated this amount of silver in water disinfection could easily be tolerated because the total absorbed dose would only be half of the NOAEL after 70 years.

Silver has been used as an integral part of EPA- and National Sanitation Foundation (NSF)-approved point-of-use (POU) water filters to prevent bacterial growth. Home water purification units (e.g., faucet-mounted devices and water pitchers) in the United States contain silverized activated carbon filters along with ion-exchange resins (Gupta et al. 1998). Today, some 50 million consumers obtain drinking water from POU devices that utilize silver (Water Quality Association 2001). These products leach silver at low levels (1–50 ppb) with no known observable adverse health effects. Such filters have been shown to prevent the growth of *Pseudomonas fluorescens* and *Pseudomonas aeruginosa* in water supplies (Russell and Hugo 1994); however, several studies have raised questions about their efficacy (Bell 1991). Reasoner et al. (1987) established that bacterial colonization of such devices occurs within a matter of days and may result in a large number of bacteria in the product water.

B. Cooling Towers/Large Building Water Distribution Systems

Cooling towers provide cooling water for air compressors and industrial processes that generate heat (Broadbent 1993). They provide an ideal environment and a suitable balance of nutrients for microbial multiplication (Martinez et al. 2004). Chlorine is a popular method for controlling such bacterial growth, but there are difficulties in maintaining disinfection efficacy, particularly at a high temperature or pH (Kim et al. 2004). Chlorination can also cause corrosion of cooling tower facilities (Kim et al. 2004).

Ag/Cu ionization has been used in cooling towers to control bacterial growth (Lin et al. 2002). In a study by Martinez et al. (2004), an appreciably reduced chlorine concentration of 0.3 parts per million (ppm or mg/L) was combined with 200 ppb Ag and 1.2 ppm Cu. This method had an appreciable impact on levels of coliform bacteria, iron-related bacteria, sulfate-reducing bacteria and slime-forming bacteria in a cooling tower (Martinez et al. 2004).

Large hot water distribution systems in hospitals and hotels have also often been attributed as a source of contaminating bacteria (Kim et al. 2002). Contaminated systems are usually treated by either superheating the water with flushing of the distal sites (heat-flush), by hyperchlorination, or by installing Ag/Cu ionization units (Stout and Yu 1997). Greater bacterial reductions have been observed with Ag/Cu ionization than with the heat-flush method (Stout et al. 1998). Ag/Cu ionization is known to provide long-term control (Liu et al. 1994; Mietzner et al. 1997) and may be used in older buildings in which the pipes could be damaged by hyperchlorination (Stout and Yu 1997). Such systems are easy to install and maintain, are relatively inexpensive, and do not produce toxic by-products (Liu et al. 1994).

One microorganism that has been commonly isolated from cooling towers is *Legionella pneumophila*, the causative agent of Legionnaires' disease (Fliermans et al. 1981; Landeen et al. 1989). Many outbreaks have been linked to cooling towers (Bentham and Broadbent 1993; Brown et al. 1999; CDC 1994) and evaporative condensers (Breiman et al. 1990). *L. pneumophila* is also commonly isolated from the periphery of hot water systems in large buildings such as hospitals, hotels, and apartment buildings where temperatures tend to be lower (Zacheus and Martikainen 1994). Ag/Cu systems have been in common use in hospitals to control *Legionella* for more than a decade (Stout and Yu 2003). Mietzner et al. (1997) reported that one such ionization system maintained effective control of *L. pneumophila* for at least 22 mon. *Legionella* may develop a tolerance to silver after a period of years, requiring higher concentrations to achieve the same effect (Rohr et al. 1999).

C. Recreational Waters

Bacteria, protozoa, and viruses may occur naturally in recreational waters or be introduced into swimming pools by bathers or through faulty connections between the filtration and sewer systems (Beer et al. 1999). Species carried by bathers include the intestinal *Streptococcus faecalis* and *Escherichia coli,* as well as skin, ear, nose, and throat organisms such as *Staphylococcus aureus, Staphylococcus epidermidis, Streptococcus salivarius, Pseudomonas aeruginosa*, and *Mycobacterium marinum* (Singer 1990). Mild to serious illnesses caused by ingestion of or contact with contaminated water (Beer et al. 1999; Craun 1988) can be the result of improperly maintained pools, spas, and hot tubs (Kebabjian 1995).

In recent years, there has been a rapid increase in the number of public, semipublic, and private pools built in Europe and America. Adequate disinfection of such waters is becoming an increasingly important health issue (Singer 1990). Traditionally, chlorine-based products are used for disinfection of swimming pools (Borgmann 2003). Chlorine produces harmful DBPs caused by the halogenation of organic compounds (urine, mucus, skin particles, hair, etc.) released into the water by swimmers (Kim et al. 2002). Thus, there is also a need for alternative disinfectants for recreational waters (Yahya et al. 1992).

Silver (Ag_2SO_4) at a low concentration (10 ppb) has been shown to kill more than 99.9% of heterotrophic bacteria in swimming pools within 30 min

(Albright et al. 1972). Silver has been used commercially in pools, but it is too slow to be used as a primary disinfectant. Regulatory agencies in some countries have recommended its use only in combination with another disinfectant (Anonymous 2006). Electrolytic generation of Ag and Cu ions allows ppb concentrations to be maintained in a convenient and reproducible manner.

D. Food and Dietary Supplements

Silver has been used to treat vinegar, fruit juices, and effervescent drinks and wine (Foegeding and Busta 1991). It is also available in Mexico as colloidal silver in gelatin ('Microdyn') for use as a consumer fruit and vegetable wash and in the U.S. as an alternative health supplement or in silver citrate complexes as food additives (Silver 2003).

E. Medical Applications

Silver has been used in numerous medical applications (Hotta et al. 1998; Yoshida et al. 1999). In dentistry, silver nitrate is effective against a number of oral bacteria including gram-negative periodontal pathogens and gram-positive streptococci that cause periodontitis (Spacciapoli et al. 2001). Dental amalgams contain approximately 35% Ag(0) and 50% Hg(0). It is unclear whether sufficient Ag(0) is released and oxidized to Ag(I) to produce an antimicrobial effect; however, the release of Hg(II) selects for metal-resistant bacteria (Silver 2003). New amalgams have therefore been introduced that contain silver alone (Silver 2003).

Silver salts have traditionally been administered to the eyes of newborn infants to prevent neonatal eye infections (Isenberg 1990). Silver ions are the most commonly used topical antimicrobial agents used in burn wound care in the Western world (Poon and Burd 2004). Both silver nitrate and silver sulphadiazine have also been used as topical antiseptics for cutaneous wounds (Fox and Modak 1974; Gupta et al. 1998; Li et al. 1997; Rosenkranz and Carr 1972). A topical cream containing 1.0% silver sulphadiazine and 0.2% chlorhexidine digluconate has been marketed as Silvazine in the U.S. (Silver 2003).

Silver sulphadiazine has recently been incorporated directly into bandages used on burns and large open wounds (Furr et al. 1994; Innes et al. 2001; Silver 2003). Unlike silver nitrate, silver sulphadiazine does not react with sulfhydryl groups or proteins. Thus, its action is not diminished in the wound (Liau et al. 1997; Modak et al. 1988). Nevertheless, the silver is still the antimicrobial portion of the molecule. Two commercial silver-coated dressings (Acticoat and Silverdin) prevented muscular invasion by *P. aeruginosa* in experimental burns in rats (Ulkur et al. 2005). *P. aeruginosa* and *S. aureus* populations were similarly affected by Silverlon, an FDA-approved wound dressing (Heggers et al. 2005).

Silver has also been used to coat vascular, urinary, and peritoneal catheters (Cicalini et al. 2004; Gentry and Cope 2005), prosthetic heart valve sewing rings (Auer et al. 2001; Ionescu et al. 2003), vascular grafts, sutures, and fracture fixation devices (Blaker et al. 2005; Darouiche 1999). Plastic indwelling catheters coated with silver compounds retard the formation of microbial biofilms (Silver 2003). Manal et al. (1996) determined that the adherence of four strains of *E. coli* was decreased by 50%–99% in comparison to silicone and latex catheters. In two separate clinical studies, 10%–12% of patients with silver-treated catheters developed bacteriuria (>100 microorganisms/mL) versus 34%–37% of patients with standard Foley catheters after 3 d. The onset of bacteriuria was thus delayed in comparison to latex catheters (Liedberg et al. 1990; Lundeberg 1986). Gentry and Cope (2005) also found a 33.5% reduction in catheter-associated urinary tract infections following the introduction of silver-coated catheters.

The complex of silver with antibiotics on the surfaces of polytetrafluoroethylene vascular grafts has been examined in a number of studies. Silver increased the elution and prolonged the duration of ciprofloxacin release in one such study (Darouiche 1999).

F. Antimicrobial Surfaces/Materials

Silver may be added to polymers (Brady et al. 2003) to confer antimicrobial activity. The result is consumer products such as washing machines, refrigerators, and ice machines that have incorporated silver (http://www.agion-tech.com/CorporateOverview.pdf, retrieved May 30, 2006; http://www.samsung.com/silvercare/index.htm, retrieved May 30, 2006). Silver has been added to plastics to produce items such as public telephones and public toilets (in Japan), toys, and infant pacifiers (Silver 2003). Johnson Matthey Chemicals (UK) utilizes an inorganic composite with immobilized slow-release silver as a preservative in their cosmetics (Silver 2003). Synthetic fabrics with silver are popular in items such as sportswear, sleeping bags, bedsheets, and dishcloths (Silver 2003; Takai et al. 2002). These fabrics are believed to reduce the level of bacterial contamination and thus odors (Silver 2003).

Silver may also be added to inorganic ceramics (e.g., zirconium phosphate, zeolite) (Cowan et al. 2003; Galeano et al. 2003; Kim et al. 1998; Kim et al. 2004) that are able to trap metal ions and may then be added to other materials (e.g., paints, plastics, waxes, polyesters) to confer antimicrobial properties (Quintavalla and Vicini 2002; Takai et al. 2002). Zeolite ceramic (sodium aluminosilicate) has a porous three-dimensional crystalline structure in which ions can reside; it has a strong affinity for silver ions and can electrostatically bind up to 40% silver (wt/wt) (Kawahara et al. 2000; Uchida 1995). Zeolites act as ion exchangers, releasing silver into the environment in exchange for other cations (Hotta et al. 1998; Kawahara et al. 2000). The amount of silver released is dependent upon the concentration of cations in the environment (Kawahara et al. 2000). The bactericidal activity of Ag-zeolite appears to result from both the effect of silver ions (Matsumura et al. 2003) and the generation of reactive oxygen species, under aerated conditions, such as superoxide anions, hydroxyl radicals, hydrogen peroxide, and singlet oxygen (Inoue et al. 2002).

Studies on stainless steel surfaces coated with zeolites containing 2.5% Ag and 14% Zn ions demonstrated significant reductions in *L. pneumophila* (Rusin et al. 2003), *S. aureus* (Bright et al. 2002), *Campylobacter jejuni*, *Salmonella typhimurium*, *Listeria monocytogenes*, and *Escherichia coli* O157:H7 (Bright KR, Gerba CP, unpublished data). Vegetative cells of *Bacillus subtilis*, *B. anthracis*, and *B. cereus* were also inactivated by at least three orders of magnitude within 24 hr by a Ag/Zn-zeolite whereas *Bacillus* spores were completely resistant under the same conditions (Galeano et al. 2003).

III. Antimicrobial Efficacy

The antimicrobial effect of silver has been demonstrated in numerous and varied applications against many different types of microorganisms including bacteria, viruses, and protozoa. An overview of the available experimental data on silver disinfection is presented in Table 1.

IV. Antimicrobial Mechanisms

Proposed mechanisms of the antibacterial and antiviral actions of silver are summarized in Table 2.

A. Antibacterial Action

The antibacterial effects of silver are not completely understood. Numerous mechanisms have been proposed. Several are generally accepted:

1. Extracellular binding or precipitation of silver to bacterial cell walls and membranes (Bellantone et al. 2002; Efrima and Bronk 1998; Goddard and Bull 1989; Slawson et al. 1992). Bacterial cell walls contain negatively charged peptidoglycans that will most likely electrostatically bind some Ag^+ on their own (Thurman and Gerba 1989).
2. Energy-dependent or independent accumulation of silver inside cells (Slawson et al. 1992). Possible active uptake of silver by a transport system for an essential metal with a similar charge or ionic size (Slawson et al. 1992; Solioz and Odermatt 1995).
3. Binding of silver to cellular proteins, including enzymes (Slawson et al. 1992). Silver is known to stain proteins (Slawson et al. 1992). It binds to sulfhydryl (–SH) groups on enzymes, leading to their inactivation (Feng et al. 2000; Liau et al. 1997; Slawson et al. 1992; Thurman and Gerba 1989)

Table 1. Microorganisms for Which Silver Has been Shown to be Effective.

Organism	Treatment	Reference
Hartmannella vermiformis *Tetrahymena pyriformis*	100 ppb Ag + 1,000 ppb Cu or 500 ppb Ag + 5,000 ppb Cu	Rohr et al. 2000
Naegleria fowleri	400 ppb Cu + 40 ppb Ag or 800 ppb Cu + 80 ppb Ag, and combined with 1.0 ppm free chlorine	Cassells et al. 1995
Mouse malaria	Silver sulphadiazine	Davis and Etris 1997
SARS-coronavirus	Ag/Al_2O_3 wafers	Han et al. 2005
Coronavirus 229E (human) Feline coronavirus Feline calicivirus	Ag/Cu zeolite	Bright KR, Gerba CP, unpublished data
HIV-1 (AIDS)	1.0, 5.0, 10.0, and 20.0 ppm of Ag_4O_4	Antelman 1992
HIV-1103	Silver thiosulfate complex encapsulated in silica gel microspheres	Davis and Etris 1997
Poliovirus (type 1 Mahoney)	400 ppb Cu + 40 ppb Ag or in combination with free chlorine at 0.2 and 0.3 ppm	Yahya et al. 1992
Poliovirus (type 1 Mahoney) Papovavirus SV-40, A 426 Adenovirus (prototype 6) Vaccinia (Elstree strain)	Sanosil Super 25 (contains silver and hydrogen peroxide) at a concentration of 0.025% and 0.1%	Kadar et al. 1993
Herpes simplex type 1	0.05% Sanosil Super 25	
Herpes vesicular stomatitis	Silver sulphadiazine	Davis and Etris 1997
Bacteriophage MS-2	400 ppb Cu + 40 ppb Ag or in combination with free chlorine at 0.2 and 0.3 ppm	Yahya et al. 1992; Thurman and Gerba 1989
Saccharomyces cerevisiae	Minimum inhibitory concentration of Ag_4O_4 is 1.25 ppm	Antelman 1992
Candida albicans	Minimal inhibitory concentration of Ag_4O_4 is 2.5–5.0 ppm	Antelman 1992
Escherichia coli *Pseudomonas aeruginosa* *Staphylococcus aureus*	AgBC (bioactive glass doped with Ag_2O) at concentrations 0.05 to 0.20 mg/ml	Bellantone et al. 2002
Escherichia coli *Staphylococcus aureus*	$AgNO_3$	Feng et al. 2000
Escherichia coli	Ceramic balls coated with Ag and Cu at a concentration of 0.05 ppm Ag and 0.05 ppm Cu	Kim et al. 2004

Table 1. *Continued*

Organism	Treatment	Reference
Pseudomonas aeruginosa *Micrococcus lutena* *Staphylococcus agalactiae*	Minimum inhibitory concentration of Ag_4O_4 is 1.25–2.5 ppm	Antelman 1992
Escherichia coli *Enterobacter cloacae* *Staphylococcus pyogenes*	Minimum inhibitory concentration of Ag_4O_4 is 2.5 ppm	Antelman 1992
Bacillus subtilis *Staphylococcus aureus* *Staphylococcus faecium*	Minimum inhibitory concentration of Ag_4O_4 is 5.0 ppm	Antelman 1992
Staphylococcus epidermidis	Minimum inhibitory concentration of Ag_4O_4 is 0.625 ppm	Antelman 1992
Escherichia coli	Ag-Zeolite	Inoue et al. 2002
Pseudomonas aeruginosa *S. aureus* (MRSA) *S. aureus* (non-MRSA)	Ag/Zn zeolite	Takai et al. 2002;
Listeria monocytogenes *Escherichia coli* *Staphylococcus aureus* *Pseudomonas aeruginosa*		Cowan et al. 2003;
Bacillus anthracis *Bacillus subtilis* *Bacillus cereus*		Galeano et al. 2003
Staphylococcus aureus *Legionella pneumophila* *Escherichia coli O157:H7*	2.5% Ag / 14% Zn (wt/wt) zeolite	Bright et al. 2002; Rusin et al. 2003;
Campylobacter jejuni *Salmonella typhimurium* *Listeria monocytogenes*		Bright KR, Gerba CP, unpublished data
Streptococcus mutans *Streptococcus mitis* *Streptococcus salivarius* *Streptococcus sanguis*	Ag-Zn Zeolite and SiO_2 at ratio concentrations of 5/55, 10/50, 20/40, and 30/30 wt%	Hotta et al. 1998
S. aureus (non-MRSA) *S. aureus* (MRSA)	0.1 ml of Silvazine (1% silver sulphadiazine + 2% chlorhexidine digluconate)	George et al. 1997
Vibrio cholerae	1.0 and 2.0 ppm of Ag_4O_4; low concentration of Ag^+	Antelman 1992; Dibrov et al. 2002
Neisseria gonorrhoeae *Treponema pallida* *Trichomonas*	Treated with silver sulphadiazine	Davis and Etris 1997
Legionella pneumophila	Treated with Ag + Cu	Davis and Etris 1997

Table 2. Summary of Mechanisms of Inactivation of Bacteria and Viruses Using Silver.

Scientific observation	Reference	Type of microbe
Release of silver into the system	Slawson et al. 1990; Inoue et al. 2002	Bacteria
Oxidative destruction catalyzed by silver	Modak and Fox 1973; Richards 1981	Bacteria
Affinity for sulfhydryl groups	Davis and Etris 1997; Feng et al. 2000	Bacteria and Viruses
Targeting of Na^+-translocating NADH: ubiquinone oxidoreductase (NQR) at low concentration of Ag^+	Dibrov et al. 2002	Bacteria
Targeting of membrane proteins	Dibrov et al. 2002	Bacteria
Inhibits oxidative metabolism required by the cells	Davis and Etris 1997; Heining 1993	Bacteria
Inhibits uptake of nutrients	Slawson et al. 1990	Bacteria
Causes metabolite leakage	Slawson et al. 1990	Bacteria
Binds to DNA	Modak and Fox 1973; Richards 1981; Thurman and Gerba 1989	Bacteria and Viruses
Site-specific Fenton mechanism	Thurman and Gerba 1989; Samuni et al. 1984; Yahya et al. 1992	Viruses
Immobilization of the virus to a surface	Thurman and Gerba 1989	Viruses
Blocks or destroys host-cell receptors	Thurman and Gerba 1989	Viruses
Inactivation of the nucleic acid within the viral capsid	Thurman and Gerba 1989	Viruses

and eventually to the inactivation of the bacteria (Liau et al. 1997). Monovalent silver ions bind to these functional groups, resulting in a stable –S–Ag group that inhibits hydrogen transfer, the source of energy transfer (Davis and Etris 1997). Silver also complexes with sulfhydryl groups in the cell membrane that are components of enzymes which participate in transmembrane energy generation and electrolyte transport (Klueh et al. 2000); this may cause the formation of R–S–S–R bonds that block respiration and electron transfer (Davis and Etris 1997; Heining 1993).

4. Binding of silver to deoxyribonucleic acid (DNA) (Thurman and Gerba 1989). Silver displaces the hydrogen bonds between adjacent nitrogens of purine and pyrimidine bases (Klueh et al. 2000; Richards 1981); this may stabilize the DNA helix and prevent replication of the DNA and sub-

sequent cell division (Modak and Fox 1973; Richards 1981; Thurman and Gerba 1989).

5. Binding of silver to electron donor groups (Thurman and Gerba 1989) containing nitrogen, oxygen, and sulfur such as amines, hydroxyls, phosphates, and thiols in cells (Grier 1983; Modak and Fox 1973).

Several observations support these proposed mechanisms. Compounds with thiol groups such as sodium thiosulfate, sodium thioglycollate, and lysozymes are able to neutralize silver activity. Silver binds to the thiol groups on these compounds and is no longer able to bind to proteins (Liau et al. 1997; Richards 1981). In a study by Bellantone et al. (2002), silver was depleted from an aqueous solution over time in the presence of bacteria. This loss was assumed to be the result of silver binding to the cell wall or accumulation inside of cells. In a separate study, silver iodide inside a polymer was able to bind sulfhydryl groups on proteins on bacterial outer membranes. It was then transported intracellularly, where it accumulated until it reached a toxicity threshold, leading to bacterial death (Brady et al. 2003). Silver accumulation has been observed in nongrowing *E. coli* cells because of both binding at the surface and intracellular uptake (Ghandour et al. 1988).

Feng et al. (2000) visualized the fate and action of silver in *E. coli* and *S. aureus* by transmission electron microscopy. In both species, the cytoplasmic membrane shrank and detached from the cell wall. An electron-light region appeared in the central region that contained large amounts of phosphorous, as determined by X-ray microanalysis. It therefore likely contained highly condensed DNA molecules. Numerous electron-dense granules both surrounded the cell wall and were deposited inside cells, surrounding, but not found within, the electron-light central region. The electron-dense granules contained significant amounts of both silver and sulfur, suggesting a combination of silver and proteins. It was proposed that the cells might produce proteins that aggregate around this nuclear region to protect DNA molecules (Feng et al. 2000). A similar mechanism has been found for heat shock proteins (Nover et al. 1983). Condensed DNA is unable to replicate. No cell growth or multiplication was observed during continuous cultivation with fresh liquid nutrient medium during the course of the experiment. Proteins were inactivated after the silver treatment and the cell wall was severely damaged in some cells. The effects were milder in *S. aureus* than in *E. coli*. The thicker cell wall of the gram-positive *S. aureus* protects it to some degree from penetration of silver ions into the cytoplasm (Feng et al. 2000).

Several other potential antibacterial mechanisms have been proposed for silver in recent years. Silver collapses the proton motive force on the cell membrane (Dibrov et al. 2002; Williams et al. 1989). Dibrov et al. (2002) found that there was a total collapse of the respiration-generated transmembrane pH gradient in vesicles and also of the membrane electric potential (in the absence of added Na^+). Low concentrations of silver ions induced massive leakage of protons (H^+) through the membrane of *Vibrio cholerae*, which resulted in the complete deenergization of the cells and most likely cell death. This effect might have been the result of modified membrane proteins or a modified phospholipid bilayer (Dibrov et al. 2002). Toxicity may also cause leakage of cellular metabolites and intracellular ions such as potassium (Slawson et al. 1992).

Silver blocks the respiratory chain of bacteria in the cytochrome oxidase and NADH-succinate-dehydrogenase region (Klueh et al. 2000). One of the primary targets of Ag^+ ions is the Na^+-translocating NADH:ubiquinone oxidoreductase (NQR). Submicromolar Ag^+ ions inhibit energy-dependent Na^+ transport in membrane vesicles; this is one of the proposed mechanisms of inactivation at low Ag^+ concentrations (Dibrov et al. 2002).

Silver also inhibits the oxidation of glucose, glycerol, fumarate, succinate, D- and L-lactate, and endogenous substrates of *E. coli* cells by the inhibition of the *b* cytochromes and cytochrome *d* at the site of substrate entry into the respiratory chain and also flavoproteins in the NADH and succinate dehydrogenase regions (Bragg and Rainnie 1973). Schreurs and Rosenberg (1982) described a mechanism specifically for silver nitrate in which it inhibits the uptake of inorganic phosphate and causes efflux of accumulated phosphate; this also induces leakage of mannitol, succinate, glutamine, and proline, causing metabolite leakage (Slawson et al. 1990).

Adsorption of atomic oxygen on the surface of silver provides a reservoir of oxygen. As a result of the catalytic action of silver, oxygen is converted to active oxygen (such as hydroxyl radicals). Silver can thus catalyze the complete destructive oxidation of bacteria (Davis and Etris 1997; Yoshida et al. 1999).

B. Antiviral Action

To date, there have been no detailed studies describing the interaction between silver and viruses. Viruses that contain sulfhydryl termini may bind silver, which might affect their replication cycle (Davis and Etris 1997). One theory is that there is a site-specific Fenton mechanism in which the metal binds to a biological molecule and is reduced by superoxide radicals or other reductants and then reoxidized by hydrogen peroxide. Continous redox reactions in a cyclic manner result in damage, as radical formation occurs near the target site of the molecule (Samuni et al. 1984; Thurman and Gerba 1989; Yahya et al. 1992).

Tzagoloff and Pratt (1964) proposed that silver modifies the adsorption of viruses to cells. Thurman and Gerba (1989) suggested that the inactivation mechanism should be one that does not require a metabolic process, for example, the immobilization of the virus to a surface, the blocking or destruction of host-cell receptors, or the inactivation of the nucleic acid within the viral capsid.

C. Antiprotozoal Action

The mechanisms by which silver acts against protozoa are not presently understood; nevertheless, many of the mechanisms that have been reported for bacteria most likely play some role against protozoa as well. For instance, silver will most certainly be able to bind to proteins on the cell membrane and, if transported inside the cell, to DNA as well. Binding to DNA could prevent replication, and binding to proteins could inhibit their function. If some of these proteins are transmembrane proteins, this may also inhibit transport and nutrient uptake.

It has been reported that silver and copper inactivate *Tetrahymena pyriformis* more easily than they do amoebas (Rohr et al. 2000). *Hartmannella* is inactivated by a concentration of 100 ppm Ag and 1,000 ppm copper (Rohr et al. 2000). There are also reports of the inactivation of *Naegleria fowleri* by the use of silver, copper, and free chlorine when used in combination (Cassells et al. 1995).

V. Silver Resistance

Rusin and Gerba (2001) defined resistance as the ability of a bacterial population to grow in working concentrations of an active disinfectant. Tolerance was defined as the ability of an organism to survive short-term exposure to a disinfectant or to survive for a longer period of time than more-sensitive bacterial strains. Many papers have been published describing silver resistance that would be considered as mere tolerance following these criteria, making a thorough discussion of silver resistance somewhat problematic. For the purpose of this review, the term "resistance" includes both true silver resistance as well as silver tolerance as the terms are not always discernible based on published descriptions of empirical data.

Some bacteria appear to have natural resistance to silver (Wood 1984). Silver-resistant bacteria are usually found in areas where bacteria are regularly exposed to silver such as in hospital burn wards, hospital water distribution systems, and contaminated soil near silver mines (Silver 2003). Two proposed mechanisms of this resistance are that silver ions are excluded from the cell or mobilized outside the cell (Slawson et al. 1992). These processes are typically performed by membrane proteins that are energy dependent and function as either ATPases or chemiosmotic cation/proton antiporters (Silver 2003). Bioaccumulation or sequestration of silver, although it does exist, is not common, and its relationship to silver resistance is unclear (Silver 2003). Silver-resistant strains of *E. coli* do not accumulate intracellular silver deposits whereas sensitive strains contain dense deposits (Starodub and Trevors 1990). The gram-positive organism *Enterococcus hirae* (formerly *Streptococcus faecalis*) possesses a homeostatic mechanism to manage intracellular copper concentration via an ion pump. The *E. hirae* CopB ATPase in membrane vesicles was found to expel both

Cu^+ and Ag^+ from the cytoplasm, causing an accumulation of Cu^+ and Ag^+ inside native inside-out membrane vesicles (Solioz and Odermatt 1995).

In gram-negative bacteria, plasmid-mediated silver resistance is believed to be the most common and typically involves energy-dependent efflux of silver from the cell. Plasmid-mediated silver resistance in *Salmonella* involves a total of nine genes and is unusual in that it includes three separate types of resistance mechanisms: a periplasmic metal-binding protein (SilE) that binds silver at the cell surface, a chemiosmotic efflux pump, and an ATPase efflux pump (SilCBA and SilP) (Silver 2003). This resistance system is somewhat homologous to the plasmid-mediated *pco* copper resistance system in *E. coli* (Silver 2003).

The *agr* gene cluster (containing genes formerly named *ybdE, ylcABCD*, and *ybcZ*) encodes a silver resistance system in *E. coli* that is homologous to the central six genes (*silA* through *silS*) of the *sil* resistance system (Silver 2003). The specific mechanisms of silver resistance have been reviewed in greater detail elsewhere (Chopra 2007; Silver 2003).

VI. Synergism with Other Disinfectants

Synergy between silver ions and other antimicrobials such as potassium permanganate, potassium peroxymonosulfate (Bright KR, Gerba CP, unpublished data), hydrogen peroxide (Armon et al. 2000; Rafter et al. 1999), biguanides (Bright KR, Gerba CP, unpublished data), chlorine (Yahya et al. 1992), chlorite and chlorate (Rafter et al. 1999), and UV light (Butkus et al. 2004) has been observed by a number of investigators against a variety of microbiological species including bacteria, viruses, and oocysts (Table 3). Interestingly, metal ions in many instances enhance the effectiveness of the system well beyond that predicted by the individual components; that is, a synergistic effect is observed. It has been postulated that the oxidizer disrupts the cell wall and effects the rapid penetration of the metallic ions into the cell where irreversible precipitation of the DNA occurs (Armon et al. 2000; Straub et al. 1995; Yahya et al. 1992). Other mechanistic interpretations are, of course, possible. For instance, at higher levels of chlorine, silver is precipitated as $AgCl_2^-$ that actually increases the sensitivity of silver-sensitive bacteria (Silver 2003).

Inactivation of *L. pneumophila* by combined copper and silver has been shown to be relatively slow when compared with that of free chlorine; nonetheless, when they were included in addition to low levels of free chlorine, the inactivation rates of bacterial indicator organisms were greater than those for free chlorine alone (Landeen et al. 1989; Yahya et al. 1990). Beer et al. (1999) found that electrolytically generated copper and silver ions used in swimming pool water along with lower levels of chlorine provided control of total coliform and heterotrophic bacteria equivalent to the control provided by high levels of chlorine. Yahya et al. (1990) demonstrated that adding 400 ppb copper and 40 ppb silver to water systems containing contaminants similar to those in swimming pools allowed the

Table 3. Synergism with Other Disinfectants.

Scientific observation	Reference
Copper and silver metals are capable of inactivating poliovirus and coliphages. This effect is greatly enhanced in the presence of oxidizers.	Yahya et al. 1992
Silver significantly enhances the effectiveness of UV light against MS-2 virus.	Butkus et al. 2004
Synergistic effect between silver, copper, and free chlorine in the inactivation of *Naegleria fowleri.*	Cassells et al. 1995
Silver shown to be synergistic with chlorite, chlorate, and oxidizers (peroxymonosulfate and hydrogen peroxide).	Rafter et al. 1999
Silver is effective in preventing biofilm formation in water. This effect is enhanced in the presence of hydrogen peroxide.	Armon et al. 2000
Silver exhibits synergistic effect with potassium monoperoxysulfate against *Acinetobacter baumannii* and *Bacillus globigii* spores.	Bright KR, Gerba CP, unpublished data
Silver and copper ions shown to have synergistic effect against *Salmonella typhimurium* and *Escherichia coli.*	Bright KR, Gerba CP, unpublished data

concentration of free chlorine to be reduced at least threefold (from 0.1 to 0.3 ppm). Enhanced inactivation rates for *E. coli, S. aureus, L. pneumophila, S. faecalis* (Landeen et al. 1989; Yahya et al. 1990), and *P. aeruginosa* (Landeen et al. 1989) were also obtained when water was treated with 400 ppb copper, 40 ppb silver, and 0.2 ppm free chlorine. These studies suggest a synergistic effect upon microorganisms subjected to copper or silver ions in the presence of low levels of chlorine.

Silver has also been shown to have synergistic activity with other metal ions such as copper and zinc. In one study (Lin et al. 1998), both copper and silver ions were found to be effective in inactivating *L. pneumophila*, and the combined effect was greater than the sum of the individual effects when each was administered alone.

In two studies, silver-resistant strains of *Acinetobacter baumannii* were found to accumulate high amounts of silver, most of which was surface bound. This resistance was reduced by the purging of a plasmid (Deshpande and Chopade 1994; Shakibaie et al. 1999). In one experiment, the plasmid was successfully transferred to *E. coli* by conjugation; however, the subsequent increased silver resistance conferred to *E. coli* was the result of the efflux of silver ions from the cell rather then accumulation (Deshpande and Chopade 1994).

If the oxidizing effect of other disinfectants damages outer cellular structure, it may permit silver ions to rapidly penetrate into the cell; this may bypass the role of silver accumulation on the cell surface as a resistance mechanism.

VII. Conclusions

Both the EPA and the WHO regard silver as safe for human consumption. It does not pose a risk to human health (World Health Organization 1996) and, in contrast to numerous other commonly utilized disinfectants, is not considered a hazardous substance (Ibarluzea et al. 1998; Kim et al. 2002; World Health Organization 1996). Silver inactivates a wide variety of microorganisms such as bacteria, viruses, and protozoa, alone or in combination with other disinfectants (Cassells et al. 1995; Davis and Etris 1997; Inoue et al. 2002), although this effect is not instantaneous.

To date, the development of resistance to silver does not appear to be a concern in real-world applications. Silver has successfully been utilized for centuries (Davis and Etris 1997; Grier 1983) and is still effective against a wide variety of microorganisms (Hotta et al. 1998; Kim et al. 2004; Rohr et al. 2000; Yahya et al. 1992). Resistance does exist in certain microorganisms (Silver 2003); however, this usually occurs in environments with high silver concentrations such as those near silver mines (Silver 2003). Silver tolerance is more likely to develop under more typical circumstances and silver usages. For example, organisms found in hospital wards and hospital water distribution systems are probably only tolerant because silver has been shown to be effective in hospital water distribution systems for several years (Blanc et al. 2005; Liu et al. 1994; Rohr et al. 1999; Stout and Yu 2003).

Further research needs to be undertaken for silver to be accepted as a disinfectant in certain applications by regulatory agencies. This research should provide sufficient information to corroborate real-world observations about the efficacy of silver as a disinfectant and any potential problems related to its use such as the development of microbial resistance.

Summary

Silver has been used as an antimicrobial for thousands of years. Over the past several decades, it has been introduced into numerous new venues such as in the treatment of water, in dietary supplements, in medical applications, and to produce antimicrobial coatings and products. Silver is often used as an alternative disinfectant in applications in which the use of traditional disinfectants such as chlorine may result in the formation of toxic by-products or cause corrosion of surfaces. Silver has also been demonstrated to produce a synergistic effect in combination with several other disinfectants. Many mechanisms of the antibacterial effect of silver have been described, but its antiviral and antiprotozoal mechanisms are not well understood. Both microbial tolerance and resistance to silver have been reported; however, the effect of silver has been observed against a wide variety of microorganisms over a period of years. Further research is needed to determine the antimicrobial efficacy of silver in these new applications and the effects of its long-term usage.

References

Albright LJ, Wentworth W, Wilson EM (1972) Technique for measuring metallic salt effects upon the indigenous heterotrophic microflora of a natural water. Water Res 6:1589–1596.

Anonymous (2006) Australian pesticides and veterinary medicines authority. http://www.apvm.gov.au/qa/poolspa_Q&A.shtml. Retrieved Feb. 1, 2006.

Antelman MS (1992) Anti-pathogenic multivalent silver molecular semiconductors. Precious Metals 16:141–149.

Armon R, Laot N, Lev O, Shuval H, Fattal B (2000) Controlling biofilm formation by hydrogen peroxide and silver combined disinfectant. Water Sci Technol 42:187–192.

Auer J, Berent R, Ng CK, Punzengruber C, Mayr H, Lassnig E, Schwarz C, Puschmann R, Hartl P, Eber B (2001) Early investigation of silver-coated Silzone heart valves prosthesis in 126 patients. J Heart Valve Dis 10:717–723.

Beer CW, Guilmartin LE, McLoughlin TF, White TJ (1999) Swimming pool disinfection. J Environ Health 61:9–13.

Bell FA (1991) Review of effects of silver impregnated carbon filters on microbial water quality. J Am Water Works Assoc 83:74–76.

Bellantone M, Williams HD, Hench LL (2002) Broad-spectrum bactericidal activity of Ag_2O-doped bioactive glass. Antimicrob Agents Chemother 46:1940–1945.

Bentham RH, Broadbent CR (1993) A model for autumn outbreaks of Legionnaires' disease associated with cooling-towers, linked to system operation and size. Epidemiol Infect 111:287–295.

Blaker JJ, Boccaccini AR, Nazhat SN (2005) Thermal characterizations of silver-containing bioactive glass-coated sutures. J Biomater Appl 20:81–98.

Blanc DS, Carrara P, Zanetti G, Francioli P (2005) Water disinfection with ozone, copper and silver ions, and temperature increase to control *Legionella:* seven years of experience in a university teaching hospital. J Hosp Infect 60:69–72.

Borgmann SR (2003) Comparative assessment of different biocides in swimming pool water. Int Biodeterior Biodegrad 51:291–297.

Brady MJ, Lisay CM, Yurkovetskiy AV, Sawan SP (2003) Persistent silver disinfectant for the environmental control of pathogenic bacteria. Am J Infect Control 31:208–214.

Bragg PD, Rainnie DJ (1973) The effect of silver ions on the respiratory chain of *Echerichia coli.* Can J Microbiol 20:883–889.

Breiman RF, Cozen W, Fields BS, Mastro TD, Carr SJ, Spika JS, Mascola L (1990) Role of air sampling in investigation of an outbreak of Legionnaires' disease associated with exposure to aerosols from an evaporative condensor. J Infect Dis 161:1257–1261.

Bright KR, Gerba CP, Rusin PA (2002) Rapid reduction of *Staphylococcus aureus* populations on stainless steel surfaces by zeolite ceramic coatings containing silver and zinc ions. J Hosp Infect 52:307–309.

Broadbent CR (1993) *Legionella* in cooling towers: practical research, design, treatment, and control guidelines. In: Barbaree JM, Breiman RF, Dufour AP (eds) *Legionella:* Current Status and Emerging Perspectives. American Society for Microbiology, Washington, DC, pp 217–222.

Brown CM, Nuorti PJ, Breiman RF, Hathcock L, Fields BS, Lipman HB, Llewellyn GC, Hofmann J, Cetron M (1999) A community outbreak of Legionnaires' disease linked to hospital cooling towers; an epidemiological method to calculate dose of exposure. Int J Epidemiol 28:353–359.

Butkus MA, Labare MP, Starke JA, Moon K, Talbot M (2004) Use of aqueous silver to enhance inactivation of coliphage MS-2 by UV disinfection. Appl Environ Microbiol 70:2848–2853.

Cassells JM, Yahya MT, Gerba CP, Rose JB (1995) Efficacy of a combined system of copper and silver and free chlorine for inactivation of *Naegleria fowleri* amoebas in water. Water Sci Technol 31:119–122.

CDC (Centers for Disease Control and Prevention) (1994) Legionnaires' disease associated with cooling-towers. MMWR 43:491–493, 499.

Chopra H (2007) The increasing use of silver-based products as antimicrobial agents: a useful development or a cause for concern? J Antimicrob Chemother Feb 19, 2007 (Epub ahead of print).

Cicalini S, Palmieri F, Petrosillo N (2004) Clinical review: new technologies for prevention of intravascular catheter-related infections. Crit Care 8:157–162.

Cowan MM, Abshire KZ, Houk SL, Evans SM (2003) Antimicrobial efficacy of a silver-zeolite matrix coating on stainless steel. J Ind Microbiol Biotechnol 30:102–106.

Craun GF (1988) Surface water supplies and health. J Am Water Works Assoc 80:40–52.

Darouiche RO (1999) Anti-infective efficacy of silver-coated medical prostheses. Clin Infect Dis 29:1371–1377.

Davis RI, Etris SF (1997) Development and functions of silver in water-purification and disease-control. Catalysis Today 36:107–114.

Deshpande LM, Chopade BA (1994) Plasmid mediated silver resistance in *Acinetobacter baumannii*. Biometals 7:49–56.

Dibrov P, Dzioba J, Gosink KK, Hase CC (2002) Chemiosmotic mechanism of antimicrobial activity of Ag^+ in *Vibrio cholerae*. Antimicrob Agents Chemother 46:2668–2670.

Efrima S, Bronk BV (1998) Silver colloids impregnating or coating bacteria. J Phys Chem B 102:5947–5950.

Environmental Protection Agency (2002) National Secondary Drinking Water Regulations. http://www.epa.gov/safewater/mcl.html. Retrieved July 7, 2006.

Feng QL, Wu J, Chen GQ, Cui FZ, Kim TN, Kim JO (2000) A mechanistic study of the antibacterial effect of silver ions on *Escherichia coli* and *Staphylococcus aureus*. J Biomed Mater Res 52:662–668.

Fliermans CB, Cherry WB, Orrison LH, Smith SJ, Tison DL, Pope DH (1981) Ecological distribution of *Legionella pneumophila*. Appl Environ Microbiol 41:9–16.

Foegeding PM, Busta FF (1991) Chemical food preservatives. In: Block SS (ed) Disinfection, Sterilization, and Preservation, 4th Ed. Lea & Febiger, Philadelphia, p 842.

Fox CL, Modak SM (1974) Mechanisms of silver sulfadiazine action on burn wound infections. Antimicrob Agents Chemother 5:582–588.

Furr JR, Russell AD, Turner TD, Andrews A (1994) Antibacterial activity of Actisorb Plus, Actisorb and silver nitrate. J Hosp Infect 27:201–208.

Galeano B, Korff E, Nicholson WL (2003) Inactivation of vegetative cells, but not spores, of *Bacillus anthracis*, *B. cereus*, and *B. subtilis* on stainless steel surfaces coated with an antimicrobial silver- and zinc-containing zeolite formulation. Appl Environ Microbiol 69:4329–4331.

Gentry H, Cope S (2005) Using silver to reduce catheter-associated urinary tract infections. Nurs Stand 19:51–54.

George N, Faoagali J, Muller M (1997) Silvazine (silver sulfadiazine and chlorhexidine) activity against 200 clinical isolates. Burns 23:493–495.

Ghandour W, Hubbard JA, Diestrung J, Hughes MN, Poole PK (1988) The uptake of silver ions by *Escherichia coli* K12: toxic effects and interaction with copper ions. Appl Microbiol Biotechnol 28:559–565.

Goddard PA, Bull TA (1989) Accumulation of silver by growing and non-growing populations of *Citrobacter intermedius* B6. Appl Microbiol Biotechnol 31:314–319.

Grier N (1983) Silver and its compounds. In: Block SS (ed) Disinfection, Sterilization, and Preservation, 3rd Ed. Lea & Febiger, Philadelphia, pp 375–389.

Gupta A, Maynes M, Silver S (1998) Effects of halides on plasmid-mediated silver resistance in *Escherichia coli*. Appl Environ Microbiol 64:5042–5045.

Han J, Duan S, Yang Q, Gao C, Zhang B, He H, Dong X (2005) Efficient and quick inactivation of SARS coronavirus and others microbes exposed to the surfaces of some metals catalysts. Biomed Environ Sci 18:176–180.

Heggers J, Goodheart RE, Washington J, McCoy L, Carino E, Dang T, Edgar P, Maness C, Chinkes D (2005) Therapeutic efficacy of three silver dressings in an infected animal model. J Burn Care Rehabil 26:53–56.

Heining CF Jr (1993) Catalyst-assisted oxidative sanitation. Ozone Sci Eng 12:533.

Hotta M, Nakamima H, Yamamoto K, Aono M (1998) Antibacterial temporary filling materials: the effect of adding various ratios of Ag-Zn-zeolite. J Oral Rehabil 25:485–489.

Ibarluzea J, Moreno B, Zigorraga C, Castilla T, Martinez M, Santamaria J (1998) Determinants of the microbiological water quality of indoor swimming-pools in relation to disinfection. Water Res 32:865–871.

Innes ME, Umraw N, Fish JS, Gomez M, Cartotto RC (2001) The use of silver coated dressings on donor site wounds: a prospective, controlled matched pair study. Burns 27:621–627.

Inoue Y, Hoshino M, Takahashi H, Noguchi T, Murata T, Kanzaki Y, Hamashima H, Sasatsu M (2002) Bactericidal activity of Ag-zeolite mediated by reactive oxygen species under aerated conditions. J Inorg Biochem 92:37–42.

Ionescu A, Payne N, Fraser AG, Giddings J, Grunkemeier GL, Butchart EG (2003) Incidence of embolism and paravalvar leak after St Jude Silzone valve implantation: experience from the Cardiff Embolic Risk Factor Study. Heart 89:1055–1061.

Isenberg SJ (1990) The dilemma of neonatal ophthalmic prophylaxis. West J Med 153:190–191.

Kadar M, Janossy L, Nagy G, Takatsy ZS, Koller M, Simon M, Pohl Ö (1993) Antiviral effect of a new disinfectant containing a silver complex and hydrogen peroxide as active agents. Wien Mitteil Wasser-Abwasser–Gewaesser 112: 62–64.

Kawahara K, Tsuruda K, Morishita M, Uchida M (2000) Antibacterial effect of silver-zeolite on oral bacteria under anaerobic conditions. Dent Mater 16:452–455.

Kebabjian RS (1995) Disinfection of public pools and management of fecal accidents. J Environ Health 58:8–12.

Kim H, Shim J, Lee S (2002) Formation of disinfection by-products in chlorinated swimming pool water. Chemosphere 46:123–130.

Kim J, Cho M, Oh B, Choi S, Yoon J (2004) Control of bacterial growth in water using synthesized inorganic disinfectant. Chemosphere 55:775–780.

Kim TN, Feng QL, Kim JO, Wu J, Wang H, Chen GC, Cui FZ (1998) Antimicrobial effects of metal ions (Ag^{+},Cu^{2+}, Zn^{2+}) in hydroxyapatite. J Mater Sci Mater Med 9:129–134.

Klueh U, Wagner V, Kelly S, Johnson A, Bryers JD (2000) Efficacy of silver-coated fabric to prevent bacterial colonization and subsequent device-based biofilm formation. J Biomed Mater Res 53:621–631.

Krause GA (1928) Neue Wege zur Wasserterilisierung. Bermann, München.

Landeen LK, Yahya MT, Gerba CP (1989) Efficacy of copper and silver ions and reduced levels of free chlorine in inactivation of *Legionella pneumophila*. Appl Environ Microbiol 55:3045–3050.

Li XZ, Nikaido H, Williams KE (1997) Silver-resistant mutants of *Escherichia coli* display active efflux of Ag^{+} and are deficient in porins. J Bacteriol 179:6127–6132.

Liau SY, Read DC, Pugh WJ, Furr JR, Russell AD (1997) Interaction of silver-nitrate with readily identifiable groups: relationship to the antibacterial action of silver ions. Lett Appl Microbiol 25:279–283.

Liedberg H, Lundeberg T, Ekman P (1990) Refinements in the coating of urethral catheters reduces the incidence of catheter-associated bacteriuria. An experimental and clinical study. Eur Urol 17:236–240.

Lin YS, Stout JE, Yu VL, Vidic RD (1998) Disinfection of water distribution systems for *Legionella*. Semin Respir Infect 13:147–159.

Lin YS, Vidic RD, Stout JE, Yu VL (2002) Negative effect of high pH on biocidal efficacy of copper and silver ions in controlling *Legionella pneumophila*. Appl Environ Microbiol 68:2711–2715.

Liu Z, Stout JE, Tedesco L, Boldin M, Hwang C, Diven WF, Yu VL (1994) Controlled evaluation of copper-silver ionization in eradicating *Legionella pneumophila* from a hospital water distribution system. J Infect Dis 169:919–922.

Lundeberg T (1986) Prevention of catheter-associated urinary-tract infections by use of silver-impregnated catheters. Lancet 2:1031.

Manal MG, Mayo MS, May LL, Simmons RB, Ahearn DG (1996) In vitro evaluation of the efficacy of a silver-coated catheter. Curr Microbiol 33:1–5.

Martinez SS, Alvarez AG, Esteban M (2004) Electrolytically generated silver and copper ions to treat cooling water: an environmentally friendly novel alternative. Int J Hydrogen Energy 29:921–932.

Matsumura Y, Yoshikata K, Kunisaki S, Tsuchido T (2003) Mode of bactericidal action of silver zeolite and its comparison with that of silver nitrate. Appl Environ Microbiol 69:4278–4281.

Mietzner S, Schwille RC, Farley A, Wald ER, Ge JH, States SJ, Libert T, Wadowsky RM, Miuetzner S (1997) Efficacy of thermal treatment and copper-silver ionization for controlling *Legionella pneumophila* in high-volume hot water plumbing systems in hospitals. Am J Infect Control 25:452–457.

Modak SM, Fox CL Jr (1973) Binding of silver sulfadiazine to the cellular components of *Pseudomonas aeruginosa*. Biochem Pharmacol 22:2391–2404.
Modak SM, Sampath L, Fox CL Jr (1988) Combined topical use of silver sulfadiazine and antibiotics as a possible solution to bacterial resistance in burn wounds. J Burn Care Rehabil 9:359–363.
Moudgal CJ, Lipscomb JC, Bruce RM (2000) Potential health effects of drinking water disinfection by-products using quantitative structure toxicity relationship. Toxicology 147:109–131.
Nover L, Scharf KD, Neuman D (1983) Formation of cytoplasmic heat shock protein granules in tomato cell cultures and leaves. Mol Cell Biol 3:1648–1655.
Pedahzur R, Katzenelson D, Barnea N, Lev O, Shuval H, Ulitzur S (2000) The efficacy of long-lasting residual drinking water disinfectants based on hydrogen peroxide and silver. Water Sci Technol 42:293–298.
Poon VK, Burd A (2004) In vitro cytotoxity of silver: implication for clinical wound care. Burns 30:140–147.
Quintavalla S, Vicini L (2002) Antimicrobial food packaging in meat industry. Meat Sci 62:373–380.
Rafter J, Grenier J, Denkewicz R (1999) US Patent #5,858,246. January 12, 1999; US Patent #5,935,609. August 10, 1999.
Reasoner DJ, Blannon JC, Geldreich EE (1987) Microbial characteristics of third faucet point-of-use devices. J Am Water Works Assoc 79:60–66.
Richards RM (1981) Antimicrobial action of silver nitrate. Microbios 31:83–91.
Rohr U, Senger M, Selenka F, Turley R, Wilhelm M (1999) Four years of experience with silver-copper ionization for control of *Legionella* in a German university hospital hot water plumbing system. Clin Infect Dis 29:1507–1511.
Rohr U, Weber S, Selenka F, Wilhelm M (2000) Impact of silver and copper on the survival of amoebae and ciliated protozoa in vitro. Int J Hyg Environ Health 203:87–89.
Rosenkranz HS, Carr HS (1972) Silver sulfadiazine: effect on the growth and metabolism of bacteria. Antimicrob Agents Chemother 2:367–372.
Rusin P, Gerba C (2001) Association of chlorination and UV irradiation to increasing antibiotic resistance in bacteria. Rev Environ Contam Toxicol 171: 1–52.
Rusin P, Bright K, Gerba C (2003) Rapid reduction of *Legionella pneumophila* on stainless steel with zeolite coatings containing silver and zinc ions. Lett Appl Microbiol 36:69–72.
Russell AD, Hugo WB (1994) Antimicrobial activity and action of silver. Prog Med Chem 31:351–370.
Samuni A, Aronovitch J, Chevio M, Czapski G (1984) In: Rottilio G, Bannister JV (eds) Life Chemistry Reports (Supplement), 2nd Ed. Harwood Academic, New York, pp 39–47.
Schreurs WJ, Rosenberg H (1982) Effect of silver ions on transport and retention of phosphate by *Escherichia coli*. J Bacteriol 152:7–13.
Shakibaie MR, Kapadnis BP, Dhakephalker P, Chopade BA (1999) Removal of silver from photographic wastewater effluent using *Acinetobacter baumannii* BL54. Can J Microbiol 45:995–1000.
Silver S (2003) Bacterial silver resistance: molecular biology and uses and misuses of silver compounds. FEMS Microbiol Rev 27:341–353.
Singer M (1990) The role of antimicrobial agents in swimming pools. Int Biodeterior 26:159–168.

Slawson RM, Lee H, Trevors JT (1990) Bacterial interactions with silver. Biol Metals 3:151–154.
Slawson RM, Van Dyke MI, Lee H, Trevors JT (1992) Germanium and silver resistance, accumulation, and toxicity in microorganisms. Plasmid 27:72–79.
Solioz M, Odermatt A (1995) Copper and silver transport by CopB-ATPase in membrane vesicles of *Enterococcus hirae*. J Biol Chem 270:9217–9221.
Spacciapoli P, Buxton D, Rothstein D, Friden P (2001) Antimicrobial activity of silver nitrate against periodontal pathogens. J Periodontal Res 36:108–113.
Starodub ME, Trevors JT (1990) Silver accumulation and resistance in *Escherichia coli* R1. J Inorg Biochem 39:317–325.
Stout JE, Yu VL (1997) Legionellosis. N Engl J Med 337:682–687.
Stout JE, Yu VL (2003) Experiences of the first 16 hospitals using copper-silver ionization for *Legionella* control: implications for the evaluation of other disinfection modalities. Infect Control Hosp Epidemiol 24:563–568.
Stout JE, Lin YS, Goetz AM, Muder RR (1998) Controlling *Legionella* in hospital water systems: experience with the superheat-and-flush method and copper-silver ionization. Infect Control Hosp Epidemiol 19:911–914.
Straub TM, Gerba CP, Zhou X, Price R, Yahya MT (1995) Synergistic inactivaction of *Escherichia coli* and MS-2 coliphage by chloramine and cupric chloride. Water Res 29:811–818.
Takai KT, Ohtsuka T, Senda Y, Nakao M, Yamamoto K, Matsuoka-Junji J, Hirai Y (2002) Antibacterial properties of antimicrobial-finished textile products. Microbiol Immunol 46:75–81.
Thurman RB, Gerba CP (1989) The molecular mechanisms of copper and silver ion disinfection of bacteria and viruses. CRC Crit Rev Environ Control 18:295–315.
Tzagoloff H, Pratt D (1964) The initial steps in infection with coliphage M13. Virology 24:372–380.
Uchida M (1995) Antimicrobial zeolite and its application. Chem Ind 46:48–54.
Ulkur E, Oncul O, Karagoz H, Celikoz B, Cavuslu S (2005) Comparison of silver-coated dressing (Acticoat), chlorhexidine acetate 0.5% (Bactigrass), and silver sulfadiazine 1% (Silverdin) for topical antibacterial effect in *Pseudomonas aeruginosa*-contaminated, full-skin thickness burn wounds in rats. J Burn Care Rehabil 26:430–433.
Von Gunten U, Driedger A, Gallard H, Salhi E (2001) By-products formation during drinking water disinfection: a tool to assess disinfection efficiency? Water Res 35:2095–2099.
Wahlberg V (1982) Reconsideration of Credé prophylaxis. A study of maternity and neonatal care. Acta Paediatr Scand Suppl 295:1–73.
Water Quality Association (2001) Use/Purchase of Home Water Treatment Systems. National Consumer Water Quality Survey, Naperville, IL.
Williams RL, Grashoff GJ, Williams DF (1989) The biocompatibility of silver. Crit Rev Biocompat 5:221–243.
Wood JM (1984) Microbiological strategies in resistance to metal ion toxicity. In: Sigel H (ed) Metal Ions in Biological Systems, vol 18. Marcel Dekker, New York, pp 333–351.
World Health Organization (1996) Guidelines for Drinking-Water Quality, 2nd Ed. WHO, Geneva, Switzerland.

Yahya MT, Landeen LK, Messina MC, Kutz SM, Schulze R, Gerba CP (1990) Disinfection of bacteria in water systems by using electrolytically generated copper: silver and reduced levels of free chlorine. Can J Microbiol 36:109–116.

Yahya MT, Straub TM, Gerba CP (1992) Inactivation of coliphage MS-2 and poliovirus by copper, silver and chlorine. Can J Microbiol 38:430–435.

Yoshida K, Tanagawa M, Matsumoto S, Yamada T, Atsuta M (1999) Antibacterial activity of resin composites with silver-containing materials. Eur J Oral Sci 107:290–296.

Zacheus OM, Martikainen PJ (1994) Occurrence of Legionellae in hot water distribution systems of Finnish apartment buildings. Can J Microbiol 40:993–999.

Manuscript received July 17; accepted July 19, 2006.

Rev Environ Contam Toxicol 191:47–89

Impact of Soil Properties on Critical Concentrations of Cadmium, Lead, Copper, Zinc, and Mercury in Soil and Soil Solution in View of Ecotoxicological Effects

Wim de Vries, Steve Lofts, Ed Tipping, Markus Meili, Jan E. Groenenberg, and Gudrun Schütze

Contents

Communicated by Pim de Voogt.

W. de Vries (✉), J.E. Groenenberg
Alterra, Wageningen University and Research Centre, Droevendaalse steeg 4, Atlas 104, P.O. Box 47, NL-6700 AA Wageningen, The Netherlands

S. Lofts, E. Tipping
Centre for Ecology and Hydrology (Lancaster), Library Avenue, Bailrigg, Lancaster LA1 4AP, United Kingdom

M. Meili
Stockholm University, Department of Applied Environmental Science (ITM), SE-10691 Stockholm, Sweden

G. Schütze
ÖKO-DATA—Ecosystem Analysis and Environmental Data Management, Hegermühlenstraße 58, 15344 Strausberg, Germany

I. Introduction

Use of Critical Metal Loads in Assessing the Risk of Metal Inputs

Concern about the input of metals to terrestrial ecosystems is related to (i) the ecotoxicological impact on soil organisms and plants (Bringmark et al. 1998; Palmborg et al. 1998) and also on aquatic organisms resulting from runoff to surface water and (ii) the uptake via food chains into animal tissues and products, which may result in health effects on animals and humans (Clark 1989). Effects on soil organisms, including microorganisms/macrofungi and soil fauna, such as nematodes and earthworms, are reduced species diversity, abundance, and biomass and changes in microbe-mediated processes (Bengtsson and Tranvik 1989; Giller et al. 1998; Vig et al. 2003). Effects on vascular plants include reduced development and growth of roots and shoots, elevated concentrations of starch and total sugar, decreased nutrient contents in foliar tissues, and decreased enzymatic activity (Prasad 1995; Das et al. 1997). A review of these phytotoxic effects is given by Balsberg-Påhlsson (1989). Effects on aquatic organisms, including algae, Crustacea, and fish, include effects on gill function (Sola et al. 1995), nervous systems (Baatrup 1991), and growth and reproduction rates (Mance 1987). Environmental quality standards or critical limits, often also denoted as Predicted No Effect Concentrations, or PNECs, for metals in soils and surface waters related to those effects serve as a guide in the environmental risk assessment process for those substances.

Following effects on soil organisms, metals may be transferred in food chains to cause effects on animals and humans, secondary poisoning. This transfer may affect (i) humans by reducing food quality of crops and animal products and (ii) affecting animal health through the accumulation in organs of cattle, birds, and mammals (secondary poisoning). Heavy metal accumulation in food chains is specifically considered important with respect to cadmium (Cd) and mercury (Hg), and to a lesser extent for lead (Pb), for all of which no biological functions are known (Clark 1989). The only exception known is that Cd appears to be essential under zinc (Zn)-limiting conditions for one marine diatom (Lee et al. 1995).

One risk assessment approach, used successfully in international negotiations on the reduction of atmospheric deposition of pollutants, is to determine the maximum load of constant atmospheric pollution that causes no or tolerable damage, "long-term acceptable load" or "critical load". A major advantage of this method is that it can be used to optimize the protection of the environment for a given international investment in pollution control by minimizing the difference between present loads and critical loads on a regional scale. A major difficulty is the quantification of the relationship between atmospheric emission, deposition, and environmental effects.

The method to calculate critical loads of metals is based on the balance of all relevant metal fluxes in and out of a considered ecosystem in a future steady-state situation. First approaches were described in manuals for calculation of critical loads of heavy metals in terrestrial ecosystems (De Vries and Bakker 1998) and aquatic ecosystems (De Vries et al. 1998). These methods were discussed at various international workshops (Gregor et al. 1997, 1999). An important development in the calculation and mapping of heavy metals was the results of a first preliminary European mapping exercise on critical loads related to ecotoxicological effects of Cd and Pb (Hettelingh et al. 2002) using a guidance document provided by De Vries et al. (2002). In the most recent manual, the critical load of a metal is simply calculated as the sum of tolerable outputs from the considered system in terms of net metal uptake by plants and metal leaching/runoff (De Vries et al. 2005). These fluxes depend on the receptor considered and the related critical limits for heavy metals.

Relevant Receptors and Related Critical Limits

With respect to risks on terrestrial ecosystems, a distinction can be made between risks/effects on the health of (i) soil organisms/processes and plants (primary ecotoxicological risks) and (ii) animals, including both domestic and wild animals and humans that use groundwater for drinking water or consume crops, meat, or fish (secondary poisoning). A description of major pathways of metals in terrestrial ecosystems, including the link with aquatic ecosystems, is given in Fig. 1.

Relevant receptors in terrestrial ecosystems, distinguishing arable land, grassland and non-agricultural land (forest, heath lands), are presented in Table 1. Possible effects on soil organisms and plants (phytotoxicity) and terrestrial fauna are of concern in all types of ecosystems. Food quality criteria are, however, of relevance for arable land and grassland (limits for animal food), whereas possible secondary poisoning effects on animals are relevant in grassland (cattle) and nonagricultural land (wild animals).

For most of the receptors or compartments indicated in Table 1, critical limits have been defined related to ecotoxicological or human toxicological risks, such as the following:

- Soil: critical limits related to effects on soil organisms (microorganisms and soil invertebrates) and plants (mg/kg).
- Plants/terrestrial fauna: critical limits in plant tissue, animal products (meat), or target organs, such as kidney, related to effects on plants and/or animals and on humans by consumption (food quality criteria) (mg/kg).
- Groundwater: critical limits in drinking water related to effects on humans by consumption (μg/L).

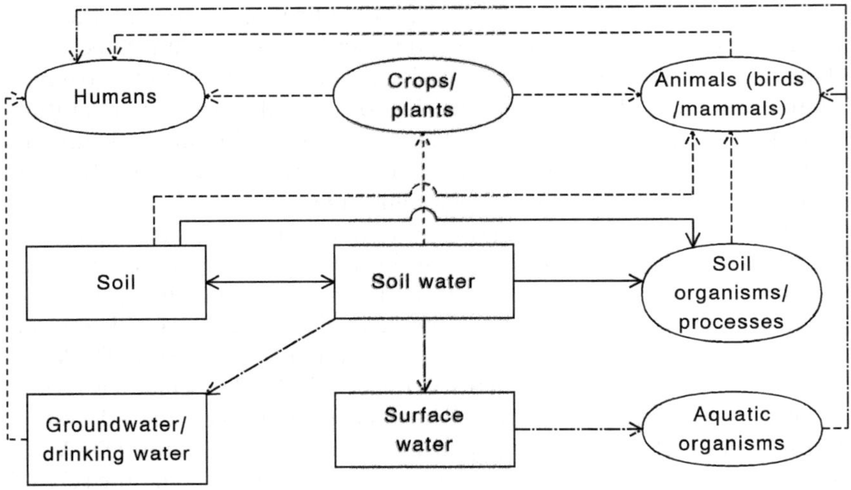

Fig. 1. Overview of the fluxes and impact pathways of metals from the soil to other compartments in terrestrial and aquatic ecosystems. *Boxes* are key "pools" and *ovals* are key "receptors." *Solid arrows* are related to fluxes/impacts within the soil ecosystem, discussed in this paper, whereas the dotted arrows refer to impacts on the health, productivity, or food quality of animals and humans due to accumulation in food chains, discussed in the subsequent paper (De Vries et al., this volume). Even though plants are also included in ecotoxicological tests for soils, *dotted arrows* are used from soil solution to plants/crops, referring to crop uptake and subsequent impacts on food quality. The *hatched arrows* are not considered in both papers, which focus on terrestrial ecosystems, and impacts on aquatic organisms in surface water are only discussed in this paper to compare results with those obtained for soil solution.

- Humans: acceptable daily intake (ADI) (μg/kg/d). This dose is the quantity of a compound to which persons can be orally exposed, on the basis of body weight, without experiencing adverse health effects.

Critical limits related to ecotoxicological effects on soil organisms and plants are limited to soil, whereas critical limits related to human toxicological risks are mainly accounted for by food quality criteria for metals in food crops, animal products (cows/sheep), fish, and drinking water consumed by them. A final critical limit can be based on the most sensitive receptor.

Need of Critical Limits for Metals in Soil and Soil Solution as a Function of Soil Properties

In view of general risk assessment, there is a need for critical limits for metals in soil. Such limits, based on laboratory studies with plants and soil organisms, are mostly related to total metal contents, either in the humus layer or the mineral soil (Bååth 1989; Bengtsson and Tranvik 1989; Tyler

Table 1. Receptors of Concern in Three Main Types of Terrestrial Ecosystems.

Receptors of concern	Type of ecosystem: Arable land	Grassland	Non-agricultural land
Ecosystem			
Soil microorganisms	+	+	+
Soil invertebrates	+	+	+
Agricultural plants	+	+	−
Wild plants	−	−	+
Human health/animal health			
Plants			
Food crops (human health)	+	−	−
Fodder crops (animal health)	−	+	−
Groundwater[a] (human health)	+	+	+
Animals			
Cattle (human and animal health)	−	+	+
Birds/mammals (animal health)	+	+	+

[a]This refers specifically to groundwater used as drinking water.

1992; Witter 1992); the same is true for effects by secondary poisoning on terrestrial fauna (Ma and van der Voet 1993; Jongbloed et al. 1994). Furthermore, limits are still often expressed as one value for a soil or transferred by a simple weighting procedure with organic and clay content that is not based on ecotoxicological assessments.

The use of a single soil metal concentration as a critical limit for ecotoxicological effects upon soil organisms has been criticized (Allen 1993) because it does not account for observed variations in the toxicity of cationic metals among soils of differing chemistry (Spurgeon and Hopkin 1996). These variations in toxicity are believed to arise because uptake and consequent toxicity of metals for many organisms occurs via soil solution (Ritchie and Sposito 2001). Specifically, the free metal ion (FMI) in soil solution is believed to be the form that is available for interactions with organisms (Lanno et al. 1999). This rationale is based on the principle of the free ion activity model (FIAM) (Morel 1983; Campbell 1995) and the biotic ligand model (BLM) (Di Toro et al. 2001; Santore et al. 2001), as discussed in the Methods section. The evidence that soil properties such as soil organic matter content, clay content, and, specifically, pH do affect the bioavailability and toxicity of metals in biota (Spurgeon and Hopkin 1996; Van Gestel and Koolhaas 2004) is hardly accounted for in the critical limits assessment.

Because effects on microorganisms, plants, and, to a large extent, also invertebrates occur through the soil solution, in particular by the free metal ions, an approach to set critical limits for FMI is in particularly appropriate to evaluate the risks of effects. It enables the consideration of the chemistry of soils (and soil solution) and their influence on the toxicity of metals. Furthermore, in view of critical load assessments for terrestrial ecosystems, there is a need for critical metal concentrations in soil solution as the critical metal leaching rate is the most important term in deriving critical loads. Because metal concentrations in soil solution are hardly ever measured, such concentrations need to be derived from critical metal concentrations in the soil with so-called transfer functions, which relate the partitioning of free metal ion concentrations in soil solution and metal concentrations in the solid phase with soil properties.

Approach

The aim of this review is to derive critical concentrations for Cd, Pb, copper (Cu), zinc (Zn), and Hg in soil and soil solution in view of impacts on soil organisms/soil processes (solid arrows, Fig. 1) while considering the effect of soil properties. The critical concentrations include (i) reactive and total metal concentrations in soils and (ii) free metal ion concentrations and total metal concentrations in soil solution. First, we present the methodologies used to derive those concentrations based on available ecotoxicological research data on impacts on soil organisms and plants in terrestrial ecosystems. We then present results obtained for Cd, Pb, Cu, and Zn, followed by a critical evaluation of the assumptions related to the derivation and use of these critical limits. A separate section is related to Hg. In a subsequent chapter (see De Vries et al., this volume), an overview is given of critical limits of Cd, Pb, and Hg in view of the impacts on human health and on animal health caused by potential accumulation in the food chain, with a focus on food quality aspects.

II. Methodological Approach

A. General Approach

Our approach to derive critical limits for Cd, Pb, Cu, and Zn for soil and soil solution as a function of soil properties is based on the standard Organisation for Economic Co-operation and Development (OECD) approach for calculating Maximum Permissible Concentrations (MPCs) or critical limits of substances in the soil (OECD 1989). The toxicity data refer to No Observed Effects Concentration (NOECs) or Lowest Observed Effects Concentration (LOECs) for metals in soils or surface water, based on chronic toxicity tests. From a range of NOEC data, a hazardous concentration at which p% of the species in an ecosystem is potentially affected, or 100-p% is protected (HC_p) is derived from the species sensitivities distribution (SSD). In line with the OECD approach, a concentration of a certain compound was considered hazardous when the probability of selecting a species with a NOEC below this concentration equals 5%; this implies that theoretically 95% of the species within an ecosystem are protected. Using this method, the 95% protection level calculated with 50% confidence is regarded as the maximum permissible concentration ($MPC = HC_5$).

The use of critical total metal concentrations in soil solution requires NOEC data for soil solution that are either directly based on measurements or derived from NOEC soil data. Because NOEC data on free ionic Cd, Pb, Cu, and Zn concentrations in soil solution are rarely available, the derivation of critical limit functions for metals in solution was based on NOEC and EC_{10} endpoint from (i) organisms that are exposed to the metal via the soil solution (plants, microorganisms, and soft-bodied soil invertebrates), (ii) accompanied by data on soil properties (pH and organic matter content) to allow the calculation of dissolved concentrations by using transfer functions, and (iii) evaluated by a statistical approach deriving limits based on a 95% protection level. Related critical reactive soil concentrations were derived in the same procedure as a function of pH and organic matter content.

In the approach, it is assumed that apart from the hard-bodied invertebrates, where soil ingestion is the major intake route, soil solution is the major pathway for metal impacts on all soil organisms and plants. This assumption is certainly valid for plants and microorganisms and for invertebrates living in soil water, such as nematodes, but is also a reliable assumption for soft-bodied invertebrates living in soil, such as earthworms (Saxe et al. 2001). The use of transfer functions is based on the assumption that effects data from ecotoxicological investigations in the laboratory can be related to a "reactive" heavy metal concentration in the soil, as the heavy metal applied in such tests is in a readily available form.

In contrast to standard statistical extrapolation methods, used to derive an HC_p, our approach is not based on the assumption that the SSD in natural ecosystems approximates a postulated statistical frequency distribution such as a log-logistic or log-normal distribution (Aldenberg and Slob 1993; Aldenberg and Jaworska 2000). Instead, we used an alternative approach called bootstrapping, presented by Newman et al. (2000), requiring no a priori assumed statistical distribution of the data. Briefly, the dataset of toxicity endpoints was repeatedly sampled and an HC_5 for each sample taken as the 5th percentile. The number of data points sampled was the same as the number in the whole data set, but individual data points could be sampled more than once, and thus each sample was slightly different from all the others. Calculation of MPCs with different confidence levels was possible by taking different percentiles of the sample HC_5. For example, the median (50%) and 5% of all sampled HC_5 was taken to be the MPC with 50% and 95% confidence level, respectively. Critical limits

Table 2. Critical Limits for Cd and Pb for Ecotoxicológical Effects in Soils (mg/kg soil), Calculated by (a) Assuming a Log-logistic Distribution of Toxic Endpoints and (b) Assuming no Statistical Distribution of Endpoints and Calculating Critical Limits by Bootstrapping.

	Critical limits from a log-logistic distribution		Critical limits from "bootstrapping"	
Metal	50% confidence	95% confidence	50% confidence	95% confidence
Cd	4.5	2.8	3.8	2.9
Pb	63	48	75	21

for Cd and Pb using the data of Klepper and van de Meent (1997), calculated with the log-logistic and bootstrap methods, are presented in Table 2.

Next, we describe in more detail (i) the transfer functions that were used to derive critical total metal concentrations in soil and soil solution from NOEC soil data (II.B.) and (ii) the approach that was used to include impacts of soil properties on critical free, reactive, and total metal concentrations in soil and soil solution (II.C.).

B. Use of Transfer Functions to Derive Critical Free and Total Metal Concentrations

Transfer functions describing the partitioning between metals soil and soil solution were used needed to calculate critical limits for free metal ion (FMI) activities from soil toxicity data, assumed to be related to reactive soil metal contents. They were also used to recalculate critical reactive or critical total metal contents for different soil conditions. The various transfer functions used in this study are described next.

Possible Transfer Functions

Transfer functions are regression relationships that describe the partitioning of metals between soil and soil solution while accounting for the impact of soil properties. Transfer functions relating soil metal concentrations in the (soil) solid phase to soil solution refer either to the free metal ion concentration or to the total dissolved metal concentration. The latter concentration includes metals bound to inorganic complexes and dissolved organic matter but excludes metals bound to suspended particulate matter. Possibilities for the calculation of a dissolved concentration from solid-phase data are presented in Fig. 2.

Data on present metal contents are mostly (pseudo) total contents, $[M]_{tot}$, based on aqua regia destruction $[M]_{AR}$ or a concentrated nitric acid destruction $[M]_{1N\text{-}HNO_3}$. Chemically, it is the reactive metal content in soil, $[M]_{re}$, that

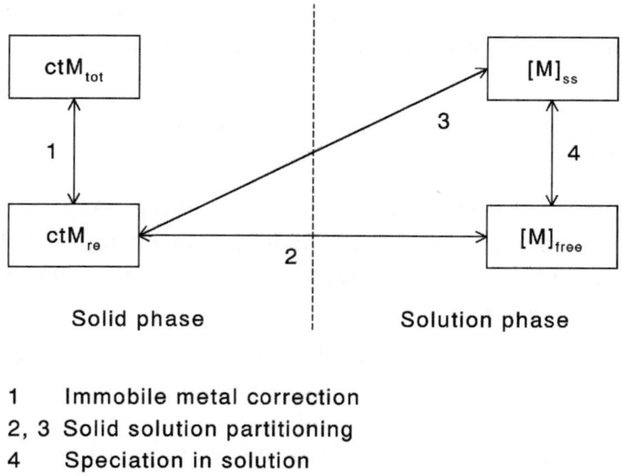

Fig. 2. Overview of relationships between metal concentrations in soil solid phase and soil solution.

interacts with the metal concentration in soil solution. Possible calculations of either a total (free and complexed) concentration of metals in solution ($[M]_{ss}$) or the free metal ion concentration or activity ($[M]_{free}$) from a total concentration in the solid phase ($[M]_{tot}$) are therefore considered inappropriate (see also Groenenberg et al. 2003). Instead, transfer functions for solid–solution partitioning can thus best be derived on the basis of reactive metal contents, based on mild HNO_3 (0.43 N), ethylene diamine tetraacetic acid (EDTA), or diethylene triamino pentaacetic acid (DTPA) extractions (Römkens et al. 2004).

In this context, we make the assumption that effects data from ecotoxicological investigations in the laboratory can also be related to a "reactive" heavy metal concentration in the soil because the heavy metal applied in such tests is in a readily available form; this implies that solid-phase transfer functions are needed to transfer reactive metal contents (NOEC data) to pseudo-total contents (relationship 1 in Fig. 2).

Regarding the soil solution, a distinction can be made between transfer functions relating free ion metal activity or concentration in solution (relationship 2 in Fig. 2) or total metal concentration in solution with reactive metal concentrations (relationship 3 in Fig. 2). Groenenberg et al. (2003) showed, for metals that form strong complexes with dissolved organic carbon (DOC), such as Pb and Cu, that the transfer functions with free ion metal activities or concentrations are always much better than the transfer functions with total concentrations. Therefore, the use of free metal activity or free metal concentration relations (relationship 2 in Fig. 2), in combination with a chemical speciation model to calculate the total dissolved metal concentration from the free metal concentration (relationship 4 in Fig. 2) is recommended. The latter aspect is discussed later.

Transfer Functions to Derive Total Soil Metal Concentrations from Reactive Soil Metal Concentrations

In this study, (pseudo)total soil metal concentrations, $[M]_{tot}$ (extracted with aqua regia), were derived from reactive metal concentration, $[M]_{re}$, assumed to be equal to added metal concentrations in laboratory toxicity data, according to the following equation:

$$\log ctM_{tot} = \beta_0 + \beta_1 \cdot \log ctM_{re} + \beta_2 \cdot \log OM + \beta_3 \cdot \log clay \quad (1)$$

where:

ctM_{tot} = the total metal content in the solid phase (mg/kg)
ctM_{re} = the reactive metal content in the solid phase (mg/kg)
OM = organic matter content in the soil (%)
clay = clay content in the soil (%)

Regression relationships were derived from a Dutch dataset containing 630 soil samples that were both extracted with 0.43 mol/L HNO_3 and aqua regia (Römkens et al. 2004). The dataset consists of a large variety of soil types with a relative wide variation in soil properties as the organic matter (median of 4% and 95% of 14%) and clay content (median of 13% and 95% of 36%). The dataset comprises both polluted and unpolluted soils. Results are shown in Table 3.

When deriving the total critical metal concentration from a critical reactive metal concentration, using Eq. 1, it should be kept in mind that the

Table 3. Values for the Coefficients β_0–β_3 in the Relationship Between Total (Aqua Regia, Being Pseudo-total) and Reactive (0.43 N HNO_3) Soil Concentrations of Cd, Pb, Cu, and Zn (Eq. 1) and the Statistical Measures R^2_{adj} and se(Y) Using a Dutch Dataset.

Metal	β_0	β_1 ctM_{re}	β_2 OM	β_3 clay	R^2_{adj}	Se(Y)[a]
Cd	0.028	0.877	0.009	0.081	0.96	0.10
Pb	0.323	0.810	0.035	0.136	0.92	0.13
Cu	0.318	0.761	0.044	0.191	0.94	0.10
Zn	0.614	0.753	−0.107	0.275	0.96	0.12

Relationships hold with ctM_{tot} and ctM_{re} in mg/kg.
[a]The standard error of the *y*-estimate on a logarithmic basis.
Source: Romkens et al. (2004).

critical soil metal concentrations are frequently higher than ambient soil concentrations, even for polluted soils. Therefore, the transfer function should preferably not be used outside its range of soil metal concentrations. The maximum values for the total (aqua regia extracted) concentrations of Cu, Zn, Cd, and Pb were approximately 330, 3,100, 40, and 1,600 mg/kg, respectively, whereas the maximum reactive (0.43 mol/L HNO_3 extracted) concentrations of Cu, Zn, Cd, and Pb were approximately 310, 2,800, 20, and 1,400 mg/kg, respectively.

Transfer Functions to Derive Dissolved Free Metal Ion Concentrations from Reactive Soil Metal Concentrations

There are various transfer functions in the literature that relate free metal ion concentrations to reactive metal contents while accounting for the effect of soil properties such as pH, organic matter content, and clay content. For example, Sauvé et al. (1997b) found that the free Cu^{2+} concentration in a sample of urban, agricultural, and forest soils could be described well as a function of pH and reactive soil Cu content alone, while Groenenberg et al. (2003) found that inclusion of both organic matter content and clay content was necessary to describe Cd^{2+} and Pb^{2+} in a dataset of Dutch soils. The transfer function used in this study was based on datasets comprising soils with a large variability in organic matter (<1% to >90%) and both the pH and organic matter content were significant variables in the transfer function that was described as follows:

$$\log\,[M]_{free} = a + b \cdot pH_{ss} + c \cdot \log OM + d \cdot \log ctM_{re} \qquad (2)$$

where:

$[M]_{free}$ = the free metal ion concentration (mol/L)
ctM_{re} = the reactive metal content in the solid phase (mol/kg)
pH_{ss} = soil solution pH.

This is entitled a c-Q relation (a relationship calculating c from Q), where c stands for the free metal ion concentration and Q stands for the reactive soil metal content. For calibration of direct transfer functions for Cd, Pb, Cu, and Zn data were drawn from seven sources:

- Sauvé et al. (1997a): Soil Pb and labile Pb in Pb-contaminated soils of various origins. Free Pb^{2+} concentrations were estimated by measurement of labile Pb using differential pulse anodic stripping voltammetry (DPASV) and speciation calculations. Metal contents in soil were determined using a concentrated HNO_3 extraction.
- Sauvé et al. (1997a): Soil Cu and free Cu^{2+} in Cu-contaminated soils of various origins: urban, forest, and agricultural. Free Cu^{2+} was measured by ion-selective electrode (ISE). Soil Cu was determined using a concentrated HNO_3 extraction.

- Sauvé et al. (2000): Soil metal and labile Cd in Cd-contaminated soils of various origins. Free Cd concentrations were estimated by measurement of labile Cd using DPASV and speciation calculations. Soil Cd was determined using a concentrated HNO_3 extraction.
- Tambasco et al. (2000): Soil Zn and free Zn^{2+} in soils of various origins: urban and forest. Free Zn^{2+} was measured using DPASV. Soil Zn was measured by extraction with 0.01 M EDTA at pH 8.6.
- Weng et al. (2001, 2002): Soil metal and free ion concentrations in sandy Dutch soils. Weng et al. (2001, 2002) measured free Cu, Zn, Cd, and Pb concentrations by the Donnan membrane technique. Metal contents in soil were determined using a 2 M HNO_3 extraction by Weng et al. (2001) and using aqua regia by weng et al. (2002).
- Tipping et al. (2003): Soil metal and free ion concentrations in UK upland soils. Free Cu, Zn, Cd, and Pb were estimated by using the WHAM6 speciation model (Tipping 1998) to speciate the soil solution. Metal contents in soil were determined using 0.43 mol/L HNO_3 extraction.

For the transfer functions derived here we have used the free ion concentration, because some of the data used (Sauvé et al. 1997a, 2000; Weng et al. 2002) express the free ion as a concentration rather than an activity. Actual differences between free activities and concentrations in soil solutions will be small compared to the expected variation in the activity or concentration with soil properties. Calculated values of the parameters in Eq. 1 are given in Table 4. According to the transfer functions, the effect of organic matter on Cu and Pb is higher than for Cd and Zn, but the effect is smaller than expected. The impact of pH and reactive metal concentration on the free metal ion concentration is, however, much higher for Cu and Pb than for Cd and Zn. More information is given by Lofts et al. (2004).

Table 4. Values for the Regression Coefficients a–d for the FMI–Reactive Metal Content Relationship (Eq. 5) for Cd, Pb, Cu, and Zn and the Statistical Measures R^2 and se(Y) Based on Results of Studies Carried out in Canada, the Netherlands, and the UK.

	a	b	c	d		
Metal		pH_{ss}	OM	log ctM_{re}	R^2	se(Y)
Cd	−0.14 (0.65)	−0.53 (0.031)	−0.60 (0.076)	0.60 (0.062)	0.62	0.53
Pb	4.33 (0.49)	−1.02 (0.032)	−0.69 (0.074)	1.05 (0.056)	0.85	0.60
Cu	4.99 (0.63)	−1.26 (0.035)	−0.63 (0.090)	0.93 (0.091)	0.90	0.61
Zn	0.55 (0.62)	−0.45 (0.027)	−0.61 (0.077)	0.57 (0.071)	0.62	0.46

Values in brackets are the standard errors for the coefficients.

C. Methodology Used to Derive Critical Metal Concentrations

Methodology to Calculate Toxic Metal Concentrations

The methodology used to calculate toxic metal concentrations was based on the evidence that toxic effects upon many soil organisms are mediated via the activity of free metals in soil solution. The principle of the free ion activity model (FIAM) is that the entry of the metal into the organism, resulting in toxicity, is considered to occur by binding to a receptor site, followed by transport into the body of the organism. If the binding step is rapid in comparison with the transport step, then the toxic effect is proportional to the amount of metal bound to the receptor, which is itself proportional to the free metal ion concentration (FMI) in bulk solution. Other cations (e.g., H^+, Na^+, Ca^{2+}) would be expected to compete with the toxic metal for the receptor site (Morel 1983). Therefore, while in a single system the toxic effect on the organism would be expected to relate to the FMI alone, when considering equivalent toxic effects across a set of systems with varying chemical composition the concentrations of these competing cations must also be considered.

Binding of the metal to the receptor is thus considered to occur in competition with other solution cations (H^+, Ca^{2+}, Mg^{2+}), so that the FMI concentration exerting a given toxic effect depends upon the concentrations of these cations in bulk solution. An increase in the concentration of any competing cation will result in an increase in the concentration of FMI required to exert a given level of toxic effect as a result of increased FMI–cation competition at the receptor site. Experimentally, this would be observed as an apparent "protective" effect of solution cations against the toxic effects of the FMI.

The approach that we used was based upon the theory of solution cations "protecting" the organism from the effects of the toxic FMI, using an empirical formulation according to Lofts et al. (2004):

$$\log{[M]_{free,toxic}} = \alpha \cdot pH_{ss} + \sum \beta_I \cdot \log{C_i} + \gamma \qquad (3)$$

where $[M]_{free,toxic}$ is the FMI concentration at the toxic endpoint, pH_{ss} is the soil solution pH, C_i is the concentration of a protecting free cation, and α, β_i, and γ are empirical coefficients. As concentrations of protecting cations such as Ca^{2+} and Mg^{2+} would be expected to covary with pH_{ss}, as a first approximation the expression may be simplified to

$$\log{[M]_{free,toxic}} = \alpha \cdot pH_{ss} + \gamma \qquad (4)$$

If toxic endpoints were available as the soil solution FMI, Eq. 4 would be directly applicable. However, literature studies routinely express the endpoint as a concentration of metal added to the soil at the start of the experiment, or as a soil metal concentration measured at the end of the experiment by chemical extraction. Therefore, Eq. 2 was used to convert the toxic soil metal concentrations to the FMI. FMI concentrations could then be calculated from the soil metal endpoints using Eq. 2 and regressed against pH_{ss} using Eq. 4. However, because pH_{ss} is a variable in Eq. 4, this approach is not statistically valid. Lofts et al. (2004) presented a methodology to calculate critical limits preserving the underlying toxicological theory of Eq. 4 while avoiding statistically invalid regression steps. At the toxic endpoint:

$$\log [M]_{free,toxic} = a + b \cdot pH_{ss} + c \cdot \log OM + d \cdot \log ctM_{re,toxic} = \alpha \cdot pH_{ss} + \gamma \quad (5)$$

Rearranging gives:

$$\log ctM_{re,toxic} + (c/d) \cdot \log OM = \phi \cdot pH_{ss} + \psi \quad (6a)$$

or

$$F = \phi \cdot pH_{ss} + \psi \quad (6b)$$

where F represents the term $\log ctM_{re,\,toxic} + (c/d) \cdot \log OM$, and ϕ and ψ are new empirical coefficients being equal to:

$$\phi = (\alpha - b)/d \quad (7a)$$

$$\psi = (\gamma - a)/d \quad (7b)$$

Note that the term [c/d] is derived from Eq. 2 and is therefore known. This expression gives the theoretical endpoint reactive soil metal, as a function of soil solution pH and % soil organic matter, and can be regressed against toxicity data to provide a function for the variation in $\log ctM_{re,toxic}$ with pH_{ss} and OM.

In literature toxicity experiments, soil pH has been estimated by chemical extraction with H_2O, KCl, or $CaCl_2$. To estimate the soil solution pH (pH_{ss}), relationships between pH by extraction with H_2O and pH_{ss}, and between pH by extraction with KCl and pH_{ss}, were established according to De Vries et al. (2005):

$$pH_{ss} = e \cdot pH_x + f \quad (8)$$

where the subscript X denotes the type of extraction used for pH (H_2O, KCl, or $CaCl_2$) and e and f are coefficients. Results are given in Table 5. For conversion from pH extracted with H_2O or with KCl, a good general relationship is found. For conversion from pH extracted with $CaCl_2$, a poorer relationship is found, but data for which such a conversion is required are uncommon in the toxicity database.

Equation 9 can be applied to give $ctM_{re,toxic}$ for a single effect on a single organism, as a function of soil solution pH and OM, by plotting F (Eq. 6b) against soil solution pH, for a series of soils of known pH and percent soil organic matter. In principle, therefore, Eq. 6 can be applied to different sets

Table 5. Results of Linear Regression Analyses of the pH in Soil Solution Against pH–H_2O and pH–KCl.

Explaining variable	N	e	f	se (pH_{ss})	R^2_{adj}
pH–H_2O	1,145	1.05	–0.28	0.45	0.84
pH–KCl	905	0.97	0.62	0.49	0.80
pH–$CaCl_2$	413	0.88	1.32	0.74	0.49

All coefficients are significant at $P > 0.999$.

of single-species data to calculate expressions for log $ctM_{re,toxic}$ for a range of pH and OM. A set of log $ctM_{re,toxic}$ values for a specific soil can then be calculated and used to define a critical limit for the soil following the methodology in Section II.A.

Methodology for Calculating Critical Limit Functions for Soil and Soil Solution

In practice, the toxicity databases for Cu, Zn, Cd, and Pb are not sufficiently comprehensive to allow for the use of the approach suggested above. Instead, a simplified approach has been used, which calculates a single critical limit function applicable to any soil. This approach centers on the theory that if the theoretical toxicity function (Eq. 6a) holds for many species and processes, then by regression of the equation against lumped multiple-endpoint data, "ecosystem average" values of the coefficients ϕ and ψ may be calculated. These coefficients then describe the apparent overall influence of soil chemistry on the endpoints. The scatter of points around the regression can be ascribed to the intrinsic variability in the sensitivity of species or processes to the toxicant, and can therefore be analyzed by a distributional approach, giving a critical limit function of the form:

$$F_{crit} = \phi \cdot pH_{ss} + \psi + \delta \tag{9}$$

where δ is a value calculated from the regression residuals for the desired level of ecosystem protection.

Lofts et al. (2004) presented in detail a method to calculate critical limit functions for Cu, Zn, Cd, and Pb using a bootstrapping technique to incorporate uncertainty in input data and parameters and to provide a convenient method to calculate a function at a given level of confidence. Briefly, a large number (10,000) samples of each toxicity dataset and of the parameters c and d (see Table 4) were taken and used to calculate 10,000 pairs of ecosystem average ϕ and ψ coefficients using Eq. 6a, and a corresponding value of δ by taking the 5th percentile of the regression residuals in *F*. These sets of coefficients were used to calculate 10,000 values of *F* (Eq. 8) at a series of pH_{ss} values and, by linear interpolation of the median *F* values, the critical limit function (with 50% confidence) was calculated:

$$F_{crit} = \log ctM_{re(crit)} + (c/d) \cdot \log OM = \phi_{crit} \cdot pH_{ss} + \psi_{crit} + \delta \quad (10)$$

where the subscript $_{CRIT}$ refers to the critical value of a variable or the values of a coefficient in the critical limit function. Equation 12 can be rearranged as

$$\log ctM_{re(crit)} = \phi_{crit} \cdot pH_{ss} - (c/d) \cdot \log OM + \psi_{crit} + \delta \quad (11)$$

The critical limit function may also be expressed in terms of the free metal ion:

$$\log [M]_{free(crit)} = \alpha_{crit} \cdot pH_{ss} + \gamma_{crit} \quad (12)$$

by calculation of critical values of α and γ according to (see Eq. 7):

$$\alpha_{crit} = b + d \cdot \phi_{crit} \quad (13a)$$

$$\gamma_{crit} = a + d \cdot (\psi_{crit} + \delta) \quad (13b)$$

Toxicity Database Used

Data sets including both NOEC or EC_{10} soil data and soil properties were used to derive NOEC soil solution data. Following procedures in the EU Risk Assessments, EC_{10} was considered equivalent to NOEC for the purposes of data gathering. NOEC and EC_{10} endpoints were used from major organisms that represent different and significant ecological functions in the ecosystem, including (i) decomposers, comprising microorganisms or microbe-mediated soil processes (e.g., enzymatic activity), (ii) consumers, such as invertebrates (earthworms and arthropods), and (iii) primary producers, specifically plants. Data for soil invertebrates were limited to soft-bodied invertebrates that are exposed to the metal via the soil solution.

To provide as far as possible consistency between the critical limits derived here and those derived under parallel EU Risk Assessment procedures for soils and surface waters, the databases used were drawn from several draft reports for these metals (EU Risk Assessment Report Cadmium, Draft Report 2003; EU Risk Assessment Report Zinc, Draft Report 2004; Environmental Risk Assessment Pb and Pb-compounds, Draft Report 2004; and Environmental Risk Assessment Cu, CuO, Cu_2O, $CuSO_4$ and $Cu_2Cl(OH)_3$, Draft report 2005). All these reports are still drafts because all limits mentioned in these reports are still under discussion by the EU. The only modification of the databases required was the removal of those endpoints for which the soil organic matter content was not provided. Metal concentrations in the control soils were not considered in deriving toxic endpoints; i.e., the added metal endpoint was used. This was the most suitable approach because the transfer functions that were applied to derive

Table 6. Numbers of Ecotoxicological Datasets Used for Copper, Zinc, Cadmium, and Lead.

Type of study	Cadmium	Lead	Copper	Zinc
Plants				
Studies	6	4	7	4
Species/groups	7	5	6	5
Endpoints	26	5	11	9
Invertebrates				
Studies	7	6	29	15
Species/groups	5	3	14[b]	7[c]
Endpoints	13	8	43	55
Microbial processes[a]				
Studies	9	14	14	10
Processes	4	10	7	7
Endpoints	18	35	33	21
Total				
Studies	22	24	50	29
Species/processes	16	18	27	19
Endpoints	57	48	87	85

[a]Respiration and substrate–induced respiration considered as one process, and all nitrogen transformation processes considered as one process.

[b]Including two community studies on nematodes and one community study on microarthropods.

[c]Including one community study on nematodes.

free metal ion concentrations are based on reactive soil metal contents. Added metal is likely to be reactive whereas some of the metal already present in the soil is likely to be in a nonbioavailable form. Furthermore, the toxic endpoint (NOEC/EC10) is always calculated by considering the effect on the organism relative to the effect in the control soil, i.e., the soil containing the background metal concentration. Thus, the effect endpoint is effectively expressed as an added metal dose, which introduces some error because of the nonlinearity of the solid–solution partitioning. If the added metal concentration is large compared to the metal already present in the soil, this error is likely to be very small. In field conditions, these limits should be considered as critical elevations above the natural background concentration. The ecotoxicological datasets used to derive critical free metal ion concentrations for Cd, Pb, Cu, and Zn as a function of pH are summarized in Table 6. The ranges in the chemical parameters in the toxicological test soils for these metals are summarized in Table 7.

Assessment of Critical Total Metal Concentrations in Soil Solution

To calculate critical loads for soils from the critical limit functions, it necessary to know the critical total metal concentration in soil drainage water,

Table 7. Ranges of Chemical Parameters in Toxicological Test Soils.

Metal	pH	OM (%)	ct $M_{soil,toxic}$ (mg/kg soil)
Cd	3.17–7.88 (6.07)	1.2–80 (4.2)	1.8–2,989 (29)
Pb	3.69–7.88 (5.99)	1.0–80 (6.2)	10–16,573 (767)
Cu	3.69–7.88 (5.99)	0.2–80 (5.0)	3.2–3,313 (120)
Zn	3.90–8.40 (5.99)	0.3–85 (5.0)	10–1,621 (158)

Values in brackets are the medians of the parameters.

$[M]_{tot,sdw(crit)}$, that corresponds to the free ion critical limit. Knowledge of $[M]_{tot,sdw\,(crit)}$ permits calculation of the leaching loss of the metal at its critical limit by combination with the leaching. Critical total metal concentrations in soil drainage water (solution and suspended particles) are determined as the sum of the critical concentration of the free metal ion M^{2+}, $[M]_{free(crit)}$, and the metals bound to (i) dissolved inorganic complexes such as MOH^+, HCO_3^+, MCl^+, and $[M]_{DIC}$; (ii) dissolved organic matter, $[M]_{DOM}$; and (iii) suspended particulate matter, $[M]_{SPM}$, according to Eq. 14:

$$[M]_{tot,sdw(crit)} = [M]_{free(crit)} + [M]_{DIC} + [M]_{DOC} \cdot [DOM] + [M]_{SPM} \cdot [SPM] \quad (14)$$

where

$[M]_{tot,sdw(crit)}$ = critical total metal concentration in soil drainage water (mg/m^3)
$[M]_{free(crit)}$ = critical free metal ion concentration (mg/m^3)
$[M]_{DIC}$ = concentration of metal bound to dissolved inorganic (carbon) species (mg/m^3)
$[M]_{DOM}$ = concentration of metal bound to dissolved organic matter in equilibrium with the critical free ion concentration (mg/kg)
[DOM] = concentration of dissolved organic matter (kg/m^3)
$[M]_{SPM}$ = concentration of metal in suspended particulate matter in equilibrium with the critical free ion concentration (mg/kg)
[SPM] = concentration of suspended particulate matter (kg/m^3)

Note that all concentrations given above refer to soil drainage water (sdw), although it has only been mentioned specifically for the total metal concentration. In soil drainage water, the concentration of suspended particulate matter is generally very small. Assuming that SPM = 0, the total metal concentration in soil drainage water is equal to the dissolved concentration

($[M]_{dis,sdw(crit)}$, being equal to the critical concentration in soil solution, $[M]_{ss(crit)}$, according to Eq. 15:

$$[M]_{ss(crit)} = [M]_{free(crit)} + [M]_{DIC} + [M]_{DOM} \cdot [DOM] \qquad (15)$$

By assuming geochemical equilibrium, the partitioning and speciation of metals over the various fractions can be calculated.

Calculating the Critical Total Metal Concentration in Soil Solution

Given the activity or free concentration of $[M]_{free}$ the concentrations of the other metal species were estimated by applying the equilibrium speciation model WHAM6 (Windermere Humic Aqueous Model, version 6; Tipping 1994, 1998). The calculation takes into account the dependence of the metal speciation on pH and competitive effects resulting from major cationic species of Mg, Al, Ca, and Fe (Tipping 2005; Tipping et al. 2002). A customized program (W6-MTC), based on WHAM 6, was used. Use of W6-MTC allows calculating critical total dissolved metal concentrations from critical pH-dependent free metal ion activities, for various combinations of pH, concentrations of soil organic matter, dissolved organic matter (DOM) or dissolved organic carbon (DOC), and suspended particulate matter (SPM) and partial CO_2 pressure (pCO_2). Calculations were made with DOC concentrations of 10, 15, 20, and 35 mg/L (DOM = 20, 35, 40, and 70 mg/L) used as average values for arable land, grassland, forest mineral topsoil (0–10 cm), and forest organic layer (O horizon), respectively, pCO_2 of 15× atmospheric value and [SPM] = 0; (De Vries et al. 2005). The calculations refer to a temperature of 10°C.

The W6-MTC program was applied by carrying out the following steps to calculate values of $[M]_{ss(crit)}$ (see Eq. 15 for its definition).

1. The concentration (g/L) of "active" fulvic acid (FA) as used in WHAM is obtained by multiplying [DOC] in mg/L by 1.3×10^{-3}. This conversion factor is based on application of the WHAM6 model to field and laboratory data for waters and soils involving Al (Tipping et al. 1991, 2002), Cu (Dwane and Tipping 1998; Vulkan et al. 2000; Bryan et al. 2002), and Cd (Tipping 2002).
2. The critical free ion concentration, $[M]_{free(crit)}$, is computed from the soil solution pH and the critical limit function (Eq. 12 with critical values of α and γ according to Table 10).
3. The activity of Al^{3+} is calculated from the pH, using equations derived by Tipping (2005). One equation applies to soils low in Al, and high in organic matter. A second equation applies to high-Al mineral soils. In the present exercise, soils with less than 20% organic matter are considered to be high in Al and those with more than 20% organic matter are considered low in Al.

4. The activity of Fe^{3+} is obtained by assuming a solubility product of $10^{2.5}$ (at 25°C) and an enthalpy of reaction of -10^7 kJ/mol (Tipping et al. 2002).

5. As a starting point, Na is assumed to be present in the soil solution at a concentration of 0.001 mol/L, balanced by equal concentrations, in equivalents, of the three major acid anions Cl^-, NO_3^-, and SO_4^{2-}. Thus, the concentrations of Cl^- and NO_3^- are each 0.000333 mol/L, whereas that of SO_4^{2-} is 0.000167 mol/L.

6. The concentration of $[M]_{free}$ and the activities of Al^{3+} and Fe^{3+} are fixed at the values obtained in steps 1–3, and the activity of H^+ is fixed from the pH. The WHAM6 model is then run to make an initial computation of inorganic solution speciation and metal binding by FA. As part of the computation, concentrations of carbonate species are obtained from pH and pCO_2. Possible metal inorganic complexes are with OH^-, Cl^-, SO_4^{2-}, HCO_3^-, and CO_3^{2-}.

7. If the result from step 6 gives an excess of positive charge, which occurs for acid solutions, the total concentrations of NO_3^- and SO_4^{2-} are increased to compensate. Then the WHAM6 program is run again.

8. If the result from step 6 gives an excess of negative charge (less acid to alkaline solutions), it is assumed that Ca provides the required additional positive charge. The WHAM6 model is run iteratively to find the total concentration of Ca that gives the correct charge balance.

9. The binding of metal to SPM is computed, by applying multiple regression equations ("transfer functions") derived for soils. The transfer function used here is Eq. 2 with the parameters mentioned in Table 4.

10. The concentrations of dissolved inorganic metal species (including $[M]_{free}$) and metal bound to dissolved organic matter, $[M]_{DOM}$, are added to obtain $[M]_{ss\ (crit)}$.

III. Derived Critical Limits for Cadmium, Lead, Copper, and Zinc Concentrations

A. Critical Limits for Reactive and Total Metal Concentrations in Soil

Critical Limits for Reactive and Total Soil Metal Concentrations

Figure 3 shows results from the toxicity data set assessment for Cd, Pb, Cu, and Zn by plotting $F(\log [M]_{re(crit)} + c/d \cdot \log [OM])$ against pH_{ss}. The regression slope is positive, indicating that $\log [M]_{re(crit)}$ increases with increasing pH_{ss}, with the exception of Cu, where the slope is not significant. The statistical significance of the slopes was tested at the 95% confidence level. Furthermore, $\log [M]_{re(crit)}$ increases with increasing soil organic matter. The relationship between $\log [M]_{re(crit)}$ and $\log [OM]$ is determined by the transfer function via the coefficient c/d. The results lead to the following relationships between the critical reactive metal concentration versus the organic

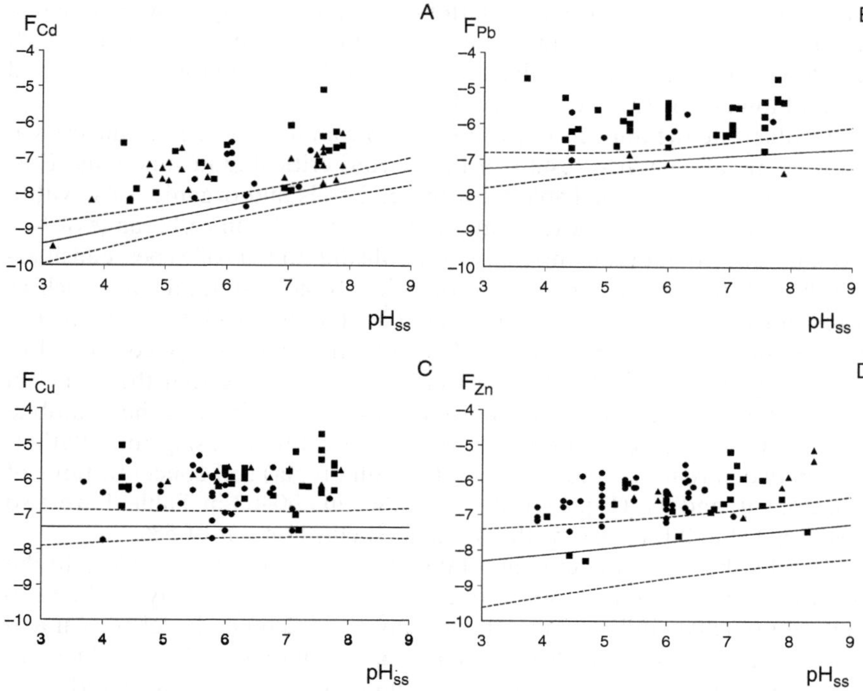

Fig. 3. Plots of F against pH_{ss} for cadmium (A), lead (B), copper (C), and zinc (D). *Triangles*, toxicity endpoints for plants; *circles*, toxicity endpoints for invertebrates; *squares*, toxicity endpoints for microbial processes. *Solid line*, median critical limit function (50% confidence); *dotted lines*, critical limit functions with 90% confidence (5% and 95%).

matter content and pH in soil solution, with the reactive metal concentrations in mol/g:

$$\log ctCd_{re(crit)} = 0.33 \cdot pH_{ss} + 1.00 \cdot \log [OM] - 10.32 \quad (16)$$

$$\log [Pb]_{re(crit)} = 0.11 \cdot pH_{ss} + 0.66 \cdot \log [OM] - 7.74 \quad (17)$$

$$\log [Cu]_{re(crit)} = 0.02 \cdot pH_{ss} + 0.68 \cdot \log [OM] - 7.54 \quad (18)$$

$$\log [Zn]_{re(crit)} = 0.14 \cdot pH_{ss} + 1.07 \cdot \log [OM] - 8.56 \quad (19)$$

The value for the dependence of the critical limits on soil organic matter, [OM], follows directly from the results of the regression coefficients c and d for the free metal ion–reactive metal content relationship (see Eq. 11 and Eq. 2 and the results for c and d in Table 4). The effect of organic matter on the critical limit for the reactive metal content is largest for Cd and Zn and lowest for Cu and Pb, whereas the reverse is true for the pH effects.

Inversely, the effects on organic matter on the partitioning between reactive metal content and free metal ion concentration are largest for Cu and Pb and lowest for Cd and Zn, whereas the pH effects are largest for Cd and Zn and lowest for Cu and Pb (see Table 4).

To illustrate the impacts of pH and organic matter content on the critical reactive metal contents, results of the various critical limit functions (Eqs. 16–19) are presented in Table 8 for mineral soils and organic soils, with a representative organic matter content of 5% and 80%, in a pH range of 4–7. Results show that it is essential to make a distinction in soil types, considering their difference in those soil properties. In acid mineral soils, such as forest topsoils, the critical reactive metal content is calculated to be nearly 10 (Cu and Pb) to more than 50 (Cd and Zn) times as low compared to highly organic soils with a high pH (see Table 8). Note that these critical limits refer to reactive metal contents, i.e., they do not include the immobile metal content. Calculated critical total metal contents, using the relationship between reactive and pseudo-total (aqua regia) soil concentrations of Cd, Pb, Cu, and Zn (Eq. 1) and the coefficients β0–β3 in Table 4 are also given in Table 8 for a soil with an assumed clay content of 5%.

Results show that the critical total metal concentrations are close to the reactive metal concentrations in organic layers. In mineral layers the ratio between reactive and total metal concentrations is much lower for Cu and Zn than for Cd and Pb. When using a clay content of 25%, values are approximately 20%–50% higher depending on the metal considered.

In comparing the results with present metal concentrations, one has to be aware that the critical concentrations are related to added metal. The values in Table 8 should thus be added to a natural background concentration before comparing them with present concentrations. Furthermore,

Table 8. Calculated Critical Reactive Metal Contents in Soil as a Function of pH and Organic Matter Content.

Metal	Organic matter content (%)	Critical reactive metal content in soil (in mg/kg)			
		pH 4	pH 5	pH 6	pH 7
Cd	5	0.56 (0.74)	1.2 (1.4)	2.6 (2.8)	5.5 (5.5)
	80	9 (9)	19 (19)	41 (41)	88 (88)
Pb	5	30 (44)	39 (54)	50 (66)	64 (81)
	80	187 (212)	241 (260)	311 (319)	400 (400)
Cu	5	6.6 (13)	6.9 (13)	7.2 (14)	7.6 (14)
	80	43 (60)	45 (63)	48 (65)	50 (67)
Zn	5	3.7 (14)	5.1 (18)	7.0 (23)	9.6 (30)
	80	71 (99)	98 (127)	135 (161)	187 (206)

Values in brackets are the critical total metal concentrations for a sandy soil with 5% clay.

there are indications that toxicity risks in the field situation are lower, thus leading to higher critical limits (see Discussion).

Normalization of Critical Limits to Organic Matter Content

The relationships derived above suggest that the critical reactive metal concentration increases with increasing soil organic matter. This relationship has important consequences for forest soils because most forest soils, at least in Northern and Central Europe, are covered by an organic layer (mor) in which many deposited pollutants are efficiently retained. Bacause plant root systems and fungi are located in this layer, there is an immediate risk of biological disturbance. Reduced decomposition of organic matter may have consequences for the mineralization of nutrients in forest soils and ultimately for forest growth.

This idea is in line with observations on the sensitivity of soil organisms in the organic layer and the mineral soil of forests. To test the relationship of the critical limit with organic matter, it was compared with field observations on the sensitivity of soil organisms to Cd, Pb, Cu, and Zn in the organic layer and the mineral soil of forests. The NOEC data for microorganisms exposed to metals in both organic layers and mineral soil compiled by Bååth (1989) were evaluated with a log-logistic fit to calculate the critical limits. Results refer to effects on enzyme synthesis and activity, litter decomposition and soil respiration. Results for Cd, Pb, Cu, and Zn are presented in Table 9. Apart from the HC_5, results are also included for HC_{20} and HC_{50} to show the impact of organic matter content on these percentiles.

The results show that the HC_5 values for the organic layer are 4 times as high for Pb compared to the mineral soil in case of the HC_5, which is comparable to results from Sweden, reporting a critical value of 34 mg/kg in the mineral soil and a range of 50–144 mg/kg for the organic layer

Table 9. Fitted Parameter Values for u and β According to Eq. (1) and Resulting Critical Limits for Total Metal Contents in Organic Layers and Mineral Soil.

						Critical limit (mg/kg)		
Metal	Layer	N^1	u	β	R^2_{adj}	HC_5	HC_{20}	HC_{50}
Cd	Organic layer	17	2.070	−0.8715	92	0.3	7.3	118
	Mineral soil	53	1.510	−0.6152	97	0.5	4.5	32
Pb	Organic layer	16	2.989	−0.2914	76	135	385	976
	Mineral soil	56	2.839	−0.4511	96	32	164	690
Cu	Organic layer	42	2.678	−0.4032	98	31	132	477
	Mineral soil	62	2.296	−0.5205	98	5.8	38	198
Zn	Organic layer	30	2.994	−0.4387	97	50	243	986
	Mineral soil	49	2.652	−0.4706	94	19	100	449

Source: Based on a compilation of NOEC data for microorganisms by Bååth.

(Bringmark, personal communication). The HC_{20} and HC_{50} are approximately 2 and 1.5 times as high. For Cd, however, the HC_5 is comparable for the organic layer and mineral soil but the HC_{20} and HC_{50} are approximately 1.5 and 4 times as high for the organic soil. For Cu and Zn, the results for the organic layer are consistently a factor 2- to 4 fold higher than for the mineral soil. Focusing on the HC_5, the results indicate comparable values for Cd in both organic layer and mineral soil, whereas for Pb, Cu, and Zn the values are 3–5 times higher for the organic layer than for the mineral soil. Results in Table 8 for 80% OM (representative for an organic layer) and 5% OM(representative for a mineral layer) at pH 4 (typical for forest soils) show that the calculated values suggest ratios near 5–7 for Pb, Cu, and Zn and even near 10 for Cd. The similar trends for the critical limit trends with the results of Bååth are a reasonable verification of the derived critical limit functions except for Cd. It should be noted that many of the effect concentrations given in Bååth (1989) are from field pollution gradients containing mixtures of metals, so it is not really possible to ascribe observed toxic effects to a single metal. These mixture effects could partly explain why the results for Cd are not comparable with those in Table 8.

B. Critical Limits for Free and Total Metal Concentrations in Soil Solution

Critical Limits for Free Metal Ion Concentrations

Figure 4 shows the Cd, Pb, Cu, and Zn toxicity dataset expressed as log $[M]_{free,toxic}$, plotted against pH. The regression slope is negative; log $[M]_{free,toxic}$ decreases with increasing pH. Results of the coefficients in the critical limit function (Eq. 12) are given in Table 10.

Lofts et al. (2004) presented critical limits based on a somewhat different set of toxicity data than that used here, which has been harmonized with the data used in the EU Risk Assessment process. The previous derived critical limit functions are shown together with the critical limit functions derived in this review in Table 10 and Fig. 4 for comparison. For Cu, Zn, and Pb, the new limit function gives higher critical free ion concentrations. Particularly for Pb the difference is appreciable, with the new function giving limits almost 1 order of magnitude (factor 10) higher at pH 3, dropping to half an order of magnitude (factor 3) at pH 8.

The new function for Cu gives limits between about 0.3 to 0.6 orders of magnitude (factor 2–4) higher, and that for Zn gives limits up to 0.3 orders (factor 2) higher. The new limit function for Cd intersects the old one at about pH 6, giving a limit 0.4 orders of magnitude lower (factor 2.5) at pH 3 and 0.2 orders (factor 1.6) higher at pH 8. The large difference in functions seen for Pb is largely caused by the removal of several sensitive endpoints relating to plant effects in tropical soils; data on such non-European soils were explicitly rejected for use under the EU Risk Assessment procedures.

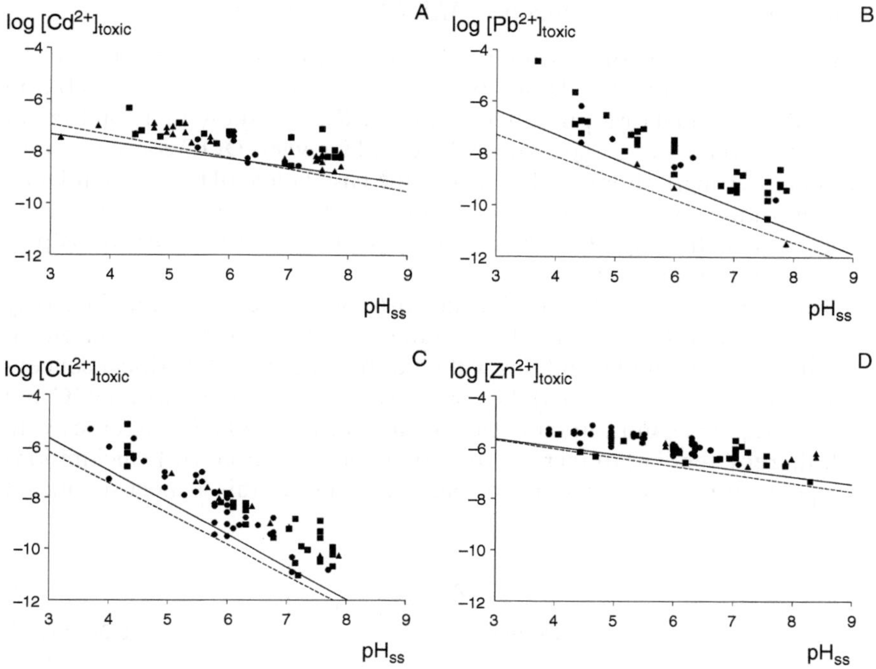

Fig. 4. Toxicity data and associated ecotoxicological critical limit functions for cadmium (A), lead (B), copper (C), and zinc (D). Data and limits are expressed as the logarithmic free metal concentration in soil solution in mg/m^3. *Triangles*, toxicity endpoints for plants; *circles*, toxicity endpoints for invertebrates; *squares*, toxicity endpoints for microbial processes. *Solid line*, median critical limit function (50% confidence); *dash-dot line*, critical limit functions previously calculated by Lofts et al. (2004).

Table 10. Coefficients in the Median Critical Limit Functions (Eq. 12) for Free Metal Ion Concentrations.

	α_{CRIT}		γ_{CRIT}	
Metal	This study	Lofts et al. (2004)	This study	Lofts et al. (2004)
Cd	–0.32	–0.43	–6.34	–5.66
Pb	–0.91	–0.83	–3.80	–4.80
Cu	–1.23	–1.21	–2.05	–2.57
Zn	–0.31	–0.34	–4.63	–4.66

Critical Limits for Total Dissolved Metal Concentrations

Critical total metal concentrations in soil solution for Cd, Pb, Cu, and Zn as a function of pH and DOC, based on calculations with the WHAM6 (W6-MTC) model, are presented in Fig. 5. Results show that total metal concentrations increase specifically below pH 5 when DOC concentrations are below 20 mg/L. For Cu, this is even the case below pH 4. At high DOC concentrations, such as in forest organic layers, the increase is generally more regular from pH 6 onward, except for Cd, where a decrease is predicted between pH 6 and 5.

The variations in the total concentrations arise from the interplay between several factors, as follow: (i) the critical free metal ion concentration decreases with pH, and this will tend to make the total critical concentration decrease with pH; (ii) the complexation of the metal with DOC and with inorganic ligands, notably carbonate species and OH^-, increases with pH, thereby tending to increase the total concentration; (iii) calcium ions compete with the toxic metals for binding by DOC; this is most significant

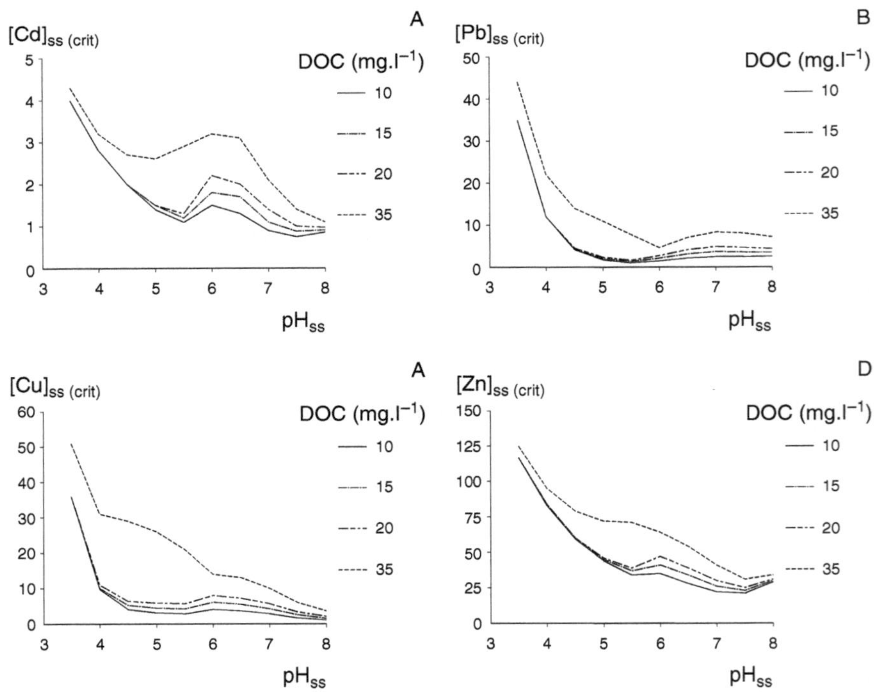

Fig. 5. Estimated total critical concentrations of cadmium (A), lead (B), copper (C), and zinc (D) in soil solution in mg/m^3 at dissolved organic carbon (DOC) concentrations of 10, 15, 20, and 35 mg/L, considered representative for arable land, grassland, forest mineral topsoil, and forest organic layer.

for the weaker binding metals (Zn, Cd) at higher pH, i.e., where the Ca concentration is increasing; and (iv) Al and Fe(III) species compete with the toxic metals for binding. The calculations assume that free concentrations of Al species are lower in organic soils than in mineral soils (Tipping 2005), and so there is stronger binding of metal by DOC and therefore the total dissolved metal concentrations are higher, in the organic horizon of the forest soil. The combination of these effects generates the complex pH dependences shown in Fig. 5.

IV. Derived Critical Limits for Mercury Concentrations

In contrast to that of Cd and Pb, the cycling of Hg in soils is controlled almost entirely by its extremely strong association with the thiols of natural organic matter (Skyllberg et al. 2003), also in soil solution where the presence of dissolved humic substances leaves virtually no free ions. The exposure of biota to Hg is thus mainly controlled by the competition between biotic and other organic ligands, and the degree of contamination of all types of natural organic matter is determined by the supply of organic matter relative to the supply of Hg at a given site (Meili 1991, 1997). Therefore, unlike the other metals, the critical limit for Hg in soils can best be set for the organically bound Hg content rather than for the free Hg ion concentration or total dissolved Hg concentration. The latter concentrations can however be derived form critical limits for mercury contents in organic soil layers, as illustrated below.

A. Critical Limits for Mercury Concentrations in Organic Soil Layers

With respect to Hg, critical soil limits presented in this study refer specifically to effects on soil microorganisms and invertebrates in the humus layer of forests, which are considered as critical receptor systems to Hg pollution (Meili et al. 2003b). Recent field studies show some observational and experimental indications of a reduced respiration in forest soils at Hg concentrations close to those encountered in rural areas of south Sweden (Bringmark and Bringmark 2001a,b). A tentative critical limit is that the mean Hg concentration in the organic top layer (O-horizon) of podzolic forest soils should not exceed the present mean level in the mentioned regions to avoid further increase. Because mercury concentrations in biologically active soils and sediments are recommended to be normalized to organic matter, OM) for the reasons mentioned above, this yields a critical limit of 0.5 mg/kg OM in these highly organic soils (Meili et al. 2003b, and references therein).

It should be stressed that this result applies to the biologically active layers of forest soils (tentatively organic matter content >10%), in which organic matter dominates the Hg cycling (transport, dilution, and toxicity). Conceptually the Hg content of total organic matter is also related (although not equal) to the Hg content in living organic matter (Meili 1997) and thus to the toxicity of soil Hg. In mineral soils where the organic matter content is generally lower than 10%, other matrices are likely to participate in controlling the cycling of Hg, and in particular the soil content of iron and aluminium (oxyhydroxides, reactive) and clay may need to be taken into account to set the critical limit. Mineral soil compartments are, however, considered less critical than the organic layer.

B. Critical Limits for Free Mercury Concentrations in Soil Solution

The concentration of free (bioreactive) Hg in soil solution can be calculated by dividing the critical concentration of "organically sorbed" Hg by the apparent distribution coefficient for Hg on (dissolved) organic matter. Field and laboratory studies, using biota to determine the concentration of bioreactive species and involving different types of soil and lake organic matter, suggest that the value for this distribution coefficient is at least 10^6 L/kg. The value seems to be fairly independent of the soil or water quality at the source from which the organic matter originates (Meili 1997). Because virtually all dissolved Hg is bound to and mobilized together with dissolved organic matter (DOM), the concentration of "free" Hg is also fairly independent of the concentration of DOM in the soil solution. Note that this distribution coefficient is operationally defined for ecotoxicological purposes by using biota to determine sum of bioavailable "free" Hg forms (which may even include organic forms), and that chemical Hg binding considered here may involve any constituent associated with natural organic matter (thiols, iron, etc.). Given the limit above for Hg bound to organic substances (0.5 mg/kg OM), the critical "free" or bioreactive Hg concentration in soil solution is below 1 ng/L, possibly even far below that.

C. Critical Limits for Total Mercury Concentrations in Soil Solution

Critical total Hg concentrations in soil solution can be calculated by using a transfer function for Hg from soil to soil solution, while assuming a similar critical Hg/org ratio in the solid phase and in the liquid phase, at least in oxic environments where binding to sulfides is negligible. The following reasoning supports this (see also Meili 1991, 1997; De Vries et al. 2003; Meili et al. 2003a; Åkerblom et al. 2004):

- As with soil solids, the Hg concentration in solution can/should be expressed on an organic matter basis, because virtually all dissolved Hg is bound to dissolved organic matter (see above). Given a typical concentration range of 10–60 mg/L for dissolved organic carbon (DOC, ≈ half of DOM) in organic forest topsoils, the distribution coefficient suggests that at least 95%–99% of all Hg is bound by organic substances if considering dissolved species alone and far more if considering the whole topsoil (>99.999%).

- If the binding properties of solid and dissolved organic matter are similar, we would expect a similar Hg/org ratio in soils and stream waters, which is indeed supported by field data (Meili 1991, 1997).
- Organic carbon concentrations in boreal stream runoff typically peaks at DOC concentrations of 15–20 mg/L; this is well within the range found in soil solutions, which supports the assumption that there are no fundamental differences between the two waters.

The critical leaching of Hg from the humus layer is related to the mobility and Hg content of dissolved organic matter because of the strong affinity of Hg for living and dead organic matter and the resulting lack of competition by inorganic ligands in this layer (Meili 1991, 1997). Therefore, Hg/OM ratios are a useful tool for calculating critical limits and loads and associated transfer functions (Meili et al. 2003b). This is the basis of the transfer function to derive total Hg concentrations in percolating topsoil solution, as follows:

$$[Hg]_{ss(crit)} = [Hg]_{[OM](crit)} \cdot f_f \cdot [DOM]_{ss} \tag{20}$$

where

$[Hg]_{ss(crit)}$ = Critical dissolved Hg concentration in soil solution (mg/m^3)
$[Hg]_{[OM](crit)}$ = Critical limit for Hg concentration in soil organic matter [OM], or the Hg/OM ratio in organic (top)soils (0.5 mg/kg OM)
f_f = Fractionation ratio, describing the Hg contamination of organic matter in solution (DOM) relative to that in solids (OM) (–)
$[DOM]_{ss}$ = Dissolved organic matter concentration in soil solution (kg/m^3)

The scale-invariant fractionation or transfer factor f_f describes the Hg partitioning between organic matter in solids and organic matter in solution and is defined as the ratio between the Hg content of DOM and that of OM (Meili et al. 2003a,b). Preliminary studies in Sweden suggest that both are of similar magnitude and that 1 may be used as a default value for f_f until deviations from unity prove to be significant (Åkerblom et al. 2004).

Based on the Hg limit of 0.5 mg/kg OM and a typical DOM concentration of 70 mg/L (or 0.07 kg/m^3; DOC = 35 mg/L) in biologically active topsoils, the critical steady-state concentration of total Hg in soil solution is 0.035 mg/m^3 or 35 ng/L (see Eq. 21). This concentration is consistent with that derived by a different approach at the watershed scale (Meili et al. 2003b) and is similar to high-end values presently observed in soil solutions and surface freshwaters (Meili 1997; Meili et al. 2003a; Åkerblom et al. 2004). Note that this ecosystem limit for soil water is much lower than the drinking water limit for Hg, but higher than that for surface freshwaters where Hg limits for fish consumption usually are exceeded at surface water concentrations of 1–5 ng/L.

V. Discussion

A. Comparison of Derived Critical Limits with Limits for Other Effects

Critical Limits for Phytotoxic Effects on Plants from NOEC Soil Solution Data

Critical limits for metals in soil solution can also be derived on the basis of NOEC soil solution data for phytotoxic effects. Results of a literature review by Lijzen et al. (2002), including data on phytotoxic effects of the metals Cd, Pb, Cu, Zn, and Hg on plants based on laboratory studies with solution culture experiments, are summarized in Table 11. A comparison of those HC_5 values with the range in critical dissolved concentrations derived for soil solution in the complete pH range (3.5–8) and DOC range (10–35 mg/L) (see Fig. 5) shows that critical concentrations of Cd are comparable in the low pH range (below 4) and even up to a pH of 7 when DOC values are high (35 mg/L), that those of Pb are comparable in a relatively high pH range (above 5) for low DOC values (up to 20 mg/L); that those of Cu are comparable in a large pH range (above 4) but only for low DOC values; and that those of Zn caused by phytotoxic effects are slightly lower than those caused by ecotoxicological effects in any pH and DOC range, but most comparable in the high pH range (above 5.5) but only at low DOC values.

In Table 11, the median and 95% range in HC_5 values in view of ecotoxicological effects are given for a pH range between 5 and 7, assuming that the phytoxicity experiments are mainly carried in this pH range. A comparison of these ranges shows that ranges are comparable. The median HC_5 values in view of ecotoxicological effects of Cd and Zn are slightly lower

Table 11. HC_5 Concentrations for Dissolved Cd, Pb, Cu, Zn, and Hg in Solution Experiments Related to Phytotoxic Impacts on Plants as Compared to Ranges in HC5 Values Derived from NOEC Data and WHAM Modelling.

Metal	No. of data	HC_5 concentration phytotoxicity (mg/m^3)	HC_5 concentration ecotoxicity (mg/m^3)
Cd	19	2.6 (0.5–07)	1.6 (1.3–3.2)
Pb	11	1.4 (0.09–7.0)	2.2 (0.5–9.2)
Cu	12	4.0 (0.3–20)	5.2 (1.7–25)
Zn	6	15 (0.3–90)	37 (18–65)
Hg	11	0.08 (0.01–0.27)	0.01–0.04

Values in brackets are the "lower limit" and "upper limit" of the 95% confidence limit.

and higher, respectively, compared to phytoxic effects (see Table 11). These results at least indicate that the modeled critical total dissolved metal concentrations (as shown in Fig. 5) are in agreement with the aforementioned literature study. For Hg, the HC_5 for phytotoxic effects is higher than the value derived from an Hg limit of 0.5 mg/kg OM critical soil. The values only overlap at the lower end of the HC_5 concentrations derived for phytotoxic impacts (Table 11).

Critical Limits for Ecotoxicological Effects on Aquatic Organisms from NOEC Surface Water Data

For aquatic ecosystems, critical limits for total dissolved metal (in mg/m^3) have been suggested on the basis of chronic toxicity data for a variety of organisms, including the major taxonomic groups, i.e., algae (unicellular and multicellular), crustacea, macrophyta, molluscs, and fish. In the effects assessment, chronic NOEC or L(E)C10 values are used rather than acute LC_{50} or EC_{50} values to derive PNEC values. As with soils, the 95% protection level calculated with 50% confidence is regarded as the MPC ($MPC = HC_5$).

A summary of effect-based critical limits, based on various EU Risk Assessment Reports for Cd, Pb, Cu, Zn, and Hg, is presented in Table 12. Values of the HC_5 are based on the 5th percentile cutoff value of various chronic toxicity data calculated with the methods described in Sub section 2.A (Aldenberg and Slob 1993; Aldenberg and Jaworska 2000). For all metals, an assessment factor was used, being a safety factor related to aspects such as the (i) endpoints covered, (ii) diversity and representativity of the taxonomic groups covered, (iii) statistical uncertainties around the 5th percentile estimate, and (iv) validation of the HC_5 with multispecies mesocosm or field data. The necessity of such a factor, varying from 1 to 4 for the various metals, can be disputed. The limits for Cd, Zn, Cu, and Pb are still under discussion by the EU.

A comparison of the critical dissolved concentrations derived for soil solution at high pH (see Fig. 5) and surface water (see Table 12) shows the critical concentration in surface waters is generally much lower for Cd and Zn, even when the assessment factor of 2 is neglected, with values in the same order of magnitude at low DOC levels; comparable for Pb when the assessment factor of 3 is neglected, but lower when included (a value near 1.6 mg/m^3 is derived for soil solution at a pH near 5.5 and a DOC concentration of 10 mg/L); comparable for Cu, for which an assessment factor of 1 is used in the official risk assessment report, and both the values of 8.2 mg/m^3 (worst case physicochemical situation) and 30.3 µg mg/m^3 (typical European physicochemical situation) are in the range encountered in soil solution between pH 4 and 8 and a DOC concentration of 10–35 mg/L; and almost equal for Hg when the assessment factor of 4 is included; a value of 0.036 mg/m^3 is also similar to high-end values presently observed in soil solutions

Table 12. Recommended Critical Limits for Dissolved Cd, Pb, Cu, Zn, and Hg Concentrations in Surface Waters, Based on Various EU Risk Assessment Reports.

Metal	Data sources	HC_5 concentration (mg/m^3)	Assessment factor	Critical limit (mg/m^3)
Cd[a]	168 single species studies 9 multi species studies	0.38	2	0.19
Pb[b]	19 freshwater NOECs/EC10s 11 saltwater NOECs/EC10s	5.0	3	1.6
Cu[c]	22 freshwater species specific NOECs/EC10s 4 multi species studies	8.2	1	8.2
Zn[d]	18 freshwater NOECs/EC10s	15.6	2	7.8
Hg[e]	30 freshwater and saltwater NOECs/EC10s	0.142	4	0.036

[a]Values based on the EU Risk Assessment Report for Cd, Draft 2003 (risk assessment cadmium metal CAS-No. 7440-43-9, EINECS-No.: 231-152-8: 2003). For Cd, a relationship with water hardness has been reported in the Report. The influence of hardness on the toxicity of cadmium can be taken into account, using three hardness classes (with hardness H in mg $CaCO_3$/L) according to 0.16 mg/m^3 if $H < 100$, 0.30 mg/m^3 if $100 < H < 200$ and 0.50 mg/m^3 if $H > 200$, when using no assessment factor.

[b]Values based on "Environmental Risk Assessment Pb and Pb-compounds—Effects Assessment to the Aquatic Compartment. Draft report 2004." Report compiled by P. Van Sprang et al.

[c]Values based on "Environmental Risk Assessment Cu, CuO, Cu_2O, $CuSO_4$, and $Cu_2Cl(OH)_3$. Effects Assessment to the Aquatic Compartment. Draft report 2005." Report Compiled by P. Van Sprang et al. The value of 8.2 mg/m^3 is based on a worst case physicochemical situation. For a typical European physicochemical situation, a value of 30.3 mg/m^3 is calculated.

[d]Values based on the EU Risk Assessment Report for Zn, Draft 2004 (Risk assessment Zinc metal CAS-No.: 7440-66-6, EINECS-No.: 231-175-3: 2004).

[e]Value based on Final Report of the Study: Identification of quality standards for priority substances in the field of water policy. Towards the Derivation of Quality Standards for Priority Substances in the Context of the Water Framework Directive (2003).

and surface freshwater (Meili 1997; Meili et al. 2003a; Åkerblom et al. 2004).

The differences observed between soil and surface water critical limits can be questioned. Analysis of aquatic ecotoxicological data by Lofts et al. (unpublished data) suggested overlap between aquatic and terrestrial toxic endpoint concentrations at a given pH. Hence, one might think of using common critical limits for both soils and freshwaters, by using the critical limit functions derived before for toxic effects on the soil ecosystem.

Table 13. Comparison of Ranges in HC5 Values for Cd, Pb, Cu, Zn, and Hg Derived from NOEC Soil Data in this study with HC_5 Concentrations Related to Phytotoxic Impacts on Plants, Impacts on Aquatic Organisms and Drinking Water Limits.

Metal	HC_5 concentration ecotoxicity (mg/m^3)	HC_5 concentration phytotoxicity (mg/m^3)	HC_5 concentration surface waters (mg/m^3)	Drinking water limit (mg/m^3)
Cd	1.6	2.6	0.19	1
Pb	2.2	1.4	1.6	10
Cu	5.2	4.0	8.2	—
Zn	37	15	7.8	—
Hg	0.02	0.08	0.036	3

However, although there is no theoretical reason why the sensitivities of soil and water organisms to metals should not be similar (assuming that uptake of the free ion from the aqueous phase is the significant mechanism leading to toxicity), this approach cannot yet be advocated because the aquatic toxicity data covered a more restricted pH range than those for the terrestrial toxicity data. More research is needed to study the possibility of using similar limits for waters and soil solution.

Overall Comparison

In Table 13, an overall comparison of median critical limits is given, including World Health Organization (WHO) data for drinking water limits. The comparison shows that the limits for ecotoxicological and phytotoxic effects are generally comparable with the HC5 for surface waters, being most stringent for Cd and Zn. The drinking water limit is comparable to these limits for Cd, but is 5 times to even 100 times higher for Pb and Hg, respectively (Table 13).

B. Uncertainties in the Calculation of Critical Limits from NOEC Data

Assumptions in Extrapolating Single-Species Toxicity Data to Ecosystem Effects

The function of risk assessment is the overall protection of the environment. Certain assumptions are made to allow extrapolation from single-species toxicity data to ecosystem effects, such as (i) ecosystem sensitivity depends on the most sensitive species and (ii) protecting ecosystem structure protects community function. It is thus assumed that protection of the most sensitive species protects ecosystem structure and function. The main motivation for introducing species sensitivity distributions (SSDs) into the MPC (critical limit) derivation is that it makes use of all available data when deriving a critical limit. The main underlying assumption of the statistical extrapolation method is that the species tested in the laboratory are a random sample of the actual species sensitivity distributions (OECD 1992).

In general, critical limits derived from extrapolations of single-species toxicological NOEC data to a maximum permissible concentration (MPC) include several uncertainties for reasons such as the following (Forbes and Forbes 1993; see also De Vries and Bakker 1998):

- Lack of representativity of the selected test species, incomparability of different endpoints
- Species sensitivities distribution not following a theoretical distribution function
- Occurrence of intra- and interspecies variations (biological variance)
- Laboratory data to field impact extrapolation, such as differences in metal availability in the laboratory and the field situation and in the exposure time
- Occurrence of intra- and interlaboratory variation of toxicity data
- Additive, synergistic, and antagonistic effects arising from the presence of other substances

In deriving critical limits in this study, we tried to overcome several of those uncertainties. First, the MPC derivation was based on a statistical extrapolation of approximately 50–90 NOECs from different species covering major taxonomic groups. Considering this amount, the data of the most sensitive endpoint might be seen as representative. Second, we did not assume an a priori theoretical distribution function. Third, we quantified the intrinsic variability in the sensitivity of species or processes to the toxicant in terms of the scatter of points around the regression described in Eq. 14. Finally, the differences between metal availability in the laboratory and field situation were accounted for in the transfer functions used. Regarding exposure time, the NOECs from chronic/long-term studies were mainly based on full-lifetime or multigeneration test studies. The other error-derived sources of scatter in the data, such as intra- and interlaboratory variation of toxicity data, cannot be assessed quantitatively.

Because of the various aforementioned uncertainties, arbitrary assessment (or safety) factors have been suggested to extrapolate from single-species laboratory data to a multispecies ecosystem, related to aspects such as the following:

- The overall quality of the database and the endpoints covered, e.g., if all the data are generated from "true" chronic studies covering all sensitive life stages
- The diversity and representativity of the taxonomic groups covered by the database, and the extent to which differences in the life forms, feeding strategies, and trophic levels of the organisms are represented

- Statistical uncertainties around the 5th percentile estimate, reflected in the goodness of fit or the size of confidence interval around the 5th percentile and consideration of different levels of confidence

The size of the assessment factor depends on the confidence with which a critical limit can be derived from the available data. This confidence increases if data are available on the toxicity to organisms at a number of trophic levels, taxonomic groups, and with lifestyles representing various feeding strategies. Thus, lower assessment factors can be used with larger and more relevant long-term data sets.

In this study no use has been made of assessment factors because sufficient NOEC data were available for major taxonomic groups to avoid the derivation of unrealistically low critical limits. This approach is supported by the Scientific Committee on Toxicity, Ecotoxicity and the Environment (CSTEE), who evaluated the risk assessment of cadmium in the framework of Council Regulation (EEC) 793/93 on the evaluation and control of risks of existing substances (CSTEE 2004).

Differences in Metal Toxicity of Laboratory-Spiked Soil and Field-Contaminated Soils

Comparison of field and laboratory NOEC soil data in the EU Risk Assessment Reports for Cu, Pb, and Zn showed consistently higher results for field data. Consequently, laboratory to field factors were used that increase the critical limit by a factor of 2 for Cu, 3 for Zn, and 4 for Pb. An important cause for this difference is the higher availability of added metals in laboratory-spiked soil compared to field-contaminated soils as a result of (i) metal-induced acidification caused by hydrolysis of the metal in solution and displacement of protons from the solid phase, (ii) higher ionic strength of the soil solution, reducing the sorption of cationic metals in soil, and (iii) the slow aging reactions that metals undergo in the field (McLaughlin et al. 2004). The increased solubility of metals in the laboratory compared was clearly demonstrated by Smolders et al. (2004), who found that Zn concentrations in soil pore water were several times higher in Zn salt-spiked samples compared to equivalent field-contaminated samples, at the same total Zn concentration. Furthermore, part of the toxicity response is likely to be from salt toxicity due to osmotic stress induced by the counterion (Cl, NO_3, SO_4), especially at high concentrations (Stevens et al. 2003). The possible decrease in pH during the experiment is not accounted for in our calculation of the dissolved metal concentration and, furthermore, the transfer function used is not derived at high ionic strengths. Only the aging factor is accounted for in our approach, by assuming that all added metal is in the reactive form, which is likely to be an overestimate. These aspects imply that the derived critical limit functions for soil and soil solution may lead to an underestimate of the critical metal concentrations.

Lack of Inclusion of Competition of Base Cations

In the approach used, the competition of "protecting" cations such as Ca^{2+} and Mg^{2+} in binding the metal to the receptor was neglected by assuming that these concentration covary with pH; this causes an uncertainty in the critical limits derived. Competition from other cations is formally considered in the biotic ligand model (BLM), which has been used to explain variability in acute toxic endpoints for several freshwater species as a function of water chemistry (Santore et al. 2001; De Schamphelaere and Janssen 2002). At present a BLM for terrestrial systems (TBLM) is under development (Thakali et al. 2005).

Uncertainties in Transfer Functions Predicting Free Metal Ion Concentrations

The transfer functions that have been used in deriving pH-dependent critical limits for the FMI in soil solution need improvements for the following reasons:

- The dataset from which these transfer functions were derived is not consistent. Metal contents in soil were derived using different extraction techniques.
- The coefficient m for the metal content in the relation for Cd is <1, which means that when the equation is written according to a Freundlich equation, $n > 1$, adsorption increases with increasing concentration.
- Critical concentrations of soil metal in the ecotoxicological experiments were higher than those used in deriving the transfer functions; this holds specifically for Cd, in which in the maximum metal content in the transfer function dataset (44 mg/kg) is much lower than in the ecotoxicological data set (2,989 mg/kg). For Pb, the difference is much less (max Pb is 14,860 mg/kg while ecotox is 16,573 mg/kg).

Another drawback is that the transfer function is based on a direct approach in which the metal concentration in soil solution is the explained variable and the soil metal content is the explaining variable, which is often referred to as c-Q relations. In this case, Q stands for the reactive metal content (ctM_{re}) and c for the dissolved (free)metal concentration in soil solution ($[M]_{free}$ or $[M]_{ss}$). Results of regression coefficients thus obtained deviate from those in which the metal content in the solid phase is the explained variable and the solution concentration is the explaining variable (together with soil properties), which is often referred to as Q-c relations. Another approach is to assume Freundlich adsorption, derive the n value by optimization, and relate Freundlich adsorption constants (K_f) to soil properties (further referred to as the K_f approach). With the K_f approach, calculations can be done in both directions. Furthermore, the parameters derived using the K_f approach are more stable with respect to the data used in the derivation (Groenenberg et al. 2003). Use of this approach is thus favorable in deriving critical limit functions.

VI. Conclusions

Despite the various uncertainties involved in the derivation, the following major conclusions can be derived from this overview chapter.

1. Critical reactive and total metal concentrations in soils should be considered as a function of soil properties, such as pH organic matter and clay content. Because these soil properties vary widely between soil types, the range in critical metal contents can be large. Comparison with present metal concentrations in the rural area shows that critical metal concentrations might be exceeded at low pH and low organic matter and clay content (acid sandy soils) because of the high metal bioavailability.
2. The derivation of a pH-dependent critical limit function for FMI activity is an appropriate tool for describing the effects of Cd, Pb, Cu, and Zn. It incorporates the effect of an increase of concentrations of competing cations in the soil solution, specifically of protons, as an increase in the concentration of the FMI required to result in a toxic effect. The pH dependence, expressed as the slope between pH and the logarithmic free metal ion concentration, is larger for Pb and Cu (slope near –1) than for Cd and Zn (slope near –0.3). The dependence of critical total metal concentrations in soil solution on pH is more complex, as the relationship is affected by DOC binding and the interaction with competing ions such as Ca, Al, and Fe(III) species. Chemical speciation models, such as WHAM, are useful to derive such relationships.
3. The FMI approach is not applicable for Hg, because nearly all Hg is bound to soil organic matter. Critical concentrations of Hg in soil solution related to effects on microbiota and invertebrates living in the humus layers of forest soils can be derived based on a limit set for the Hg content of solid organic matter and assuming a similar Hg/OM ratio in the solid phase and in the liquid phase.
4. A comparison of the critical dissolved concentrations derived for soil solution at high pH and surface water shows that critical concentrations in surface waters are generally lower for Cd and Zn but comparable for Pb, Cu, and Hg.

In summary, this review shows that critical metal concentrations in soil and soil solution related to ecotoxicological effects should be derived as a function of soil and soil solution chemistry. Most important are pH and organic matter concentrations in soil and soil solution. Future work should focus on diminishing uncertainties in the derived critical metal concentration functions by (i) further assessment of relevant NOEC data combined with soil properties, (ii) improvement and validation of transfer functions, calibrated over a range of soil metal concentrations that covers the range found in the toxic endpoint NOEC data, and (iii) including the metal that is present before the start of the experiment to allow direct comparison of the critical limits thus derived with present concentrations. Finally, most important would be to assess direct relationships between measured (free) metal ion concentrations and ecotoxicological effects.

Summary

Risk assessment for metals in terrestrial ecosystems, including assessments of critical loads, requires appropriate critical limits for metal concentrations in soil and soil solution. This chapter presents an overview of methodologies used to derive critical (i) reactive and total metal concentrations in soils and (ii) free metal ion and total metal concentrations in soil solution for Cd, Pb, Cu, Zn, and Hg, taking into account the effect of soil properties related to ecotoxicological effects. Most emphasis is given to the derivation of critical free and total metal concentrations in soil solution, using available NOEC soil data and transfer functions relating solid-phase and dissolved metal concentrations. This approach is based on the assumption that impacts on test organisms (plants, microorganisms, and soil invertebrates) are mainly related to the soil solution concentration (activity) and not to the soil solid-phase content. Critical Cd, Pb, Cu, Zn, and Hg concentrations in soil solution vary with pH and DOC level. The results obtained are generally comparable to those derived for surface waters based on impacts to aquatic organisms. Critical soil metal concentrations, related to the derived soil solution limits, can be described as a function of pH and organic matter and clay content, and varying about one order of magnitude between different soil types.

Acknowledgments

This work was funded by the Ministry of Housing, Spatial Planning and the Environment and the Ministry of Agriculture, Nature and Food Quality of the Netherlands and the UK Department of the Environment, Food and Rural Affairs under grant EPG /1/3/188.

References

Åkerblom S, Meili M, Bringmark L, Johansson K (2004) Determination of fractionation factor (f_f) in forest soil describing the Hg content of organic matter in solution relative to that in solids based on field data from Sweden. Background Document, UN/ECE-CLRTAP-ICP-Modelling and Mapping of Critical Levels and Loads and Air Pollution Effects, Risks and Trends, Workshop on Critical Loads of Heavy Metals, Berlin/Potsdam, 4–5 March 2004, http://www.icpmapping.org/html/body_ws_potsdam.html.

Aldenberg T, Jaworska JS (2000) Uncertainty of the hazardous concentration and fraction affected for normal species sensitivity distributions. Ecotoxicol Environ Saf 46:1–18.

Aldenberg T, Slob W (1993) Confidence limits for hazardous concentrations based on logistically distributed NOEC toxicity data. Ecotoxicol Environ Saf 25:48–63.

Allen HE (1993) The significance of trace metal speciation for water, sediment and soil quality standards. Sci Total Environ 134:23–45.

Bååth E (1989) Effects of heavy metals in soil on microbial processes and populations: a review. Water Air Soil Pollut 47:335–379.

Baatrup E (1991) Structural and functional-effects of heavy-metals on the nervous-system, including sense-organs, of fish. Comp Biochem Physiol 199:253–257.

Balsberg-Påhlsson AM (1989) Toxicity of heavy metals (Zn, Cu, Cd, Pb) to vascular plants. Water Air Soil Pollut 47:287–319.

Bengtsson G, Tranvik L (1989) Critical metal concentrations for forest soil invertebrates. Water Air Soil Pollut 47:381–417.

Bringmark L, Bringmark E (2001a) Lowest effect levels of lead and mercury on decomposition of mor layer samples in a long-term experiment. Water Air Soil Pollut Focus 1:425–437.

Bringmark L, Bringmark E (2001b) Soil respiration in relation to small-scale patterns of lead and mercury in mor layers of southern Swedish forest sites. Water Air Soil Pollut Focus 1:395–408.

Bringmark L, Bringmark E, Samuelsson B (1998) Effects on mor layer respiration by small experimental additions of mercury and lead. Sci Total Environ 213: 115–119.

Bryan SE, Tipping E, Hamilton-Taylor J (2002) Comparison of measured and modelled copper binding by natural organic matter in freshwaters. Comp Biochem Physiol 133:37–49.

Campbell PGC (1995) Interactions between trace metals and aquatic organisms: a critique on the free-ion activity model. In: Tessier A, Turner DR (eds) Metal Speciation and Bioavailability in Aquatic Systems. John Wiley & Sons, New York, pp 45–102.

Clark RB (1989) Metals. In: Marine Pollution. OSP, Oxford Press, pp 80–105.

CSTEE (2004) Opinion on the result of the risk assessment of: cadmium metal and cadmium oxide. Environmental part. European Commission Health & Consumer Protection Directorate General, Directorate C—Public Health and Risk Assessment C 7—Risk Assessment, Brussels.

Das P, Samantaray S, Rout GR (1997) Studies on cadmium toxicity in plants: a review. Environ Pollut 98:29–36.

De Schamphelaere KAC, Janssen CR (2002) A biotic ligand model predicting acute copper toxicity for *daphnia magna*: the effects of calcium, magnesium, sodium, potassium, and pH. Environ Sci Technol 36:48–54.

De Vries W, Bakker DJ (1998) Manual for calculating critical loads of heavy metals for terrestrial ecosystems. Guidelines for critical limits, calculation methods and input data. Report 166. DLO Winand Staring Centre, Wageningen, the Netherlands.

De Vries W, Bakker DJ, Sverdrup HU (1998) Manual for calculating critical loads of heavy metals for aquatic ecosystems. Guidelines for critical limits, calculation methods and input data. Report 165. DLO Winand Staring Centre, Wageningen, the Netherlands.

De Vries W, Schütze G, Römkens PFAM, Hettelingh, JP (2002) Guidance for the calculation of critical loads for cadmium and lead in terrestrial and aquatic ecosystems. In: Hettelingh JP, Slootweg J, Posch M, Ilyin I (eds) Preliminary

Modelling and Mapping of Critical Loads for Cadmium and Lead in Europe. RIVM Report 259101011/2002. National Institute of Public Health and Environmental Protection Bilthoven, The Netherlands, pp 17–35.

De Vries W, Schütze G, Lofts S, Meili M, Römkens PFAM, Farret R, de Temmerman L, Jakubowski M (2003) Critical limits for cadmium, lead and mercury related to ecotoxicological effects on soil organisms, aquatic organisms, plants, animals and humans. In: Schütze G, Lorent U, Spranger T (eds) Proceedings of the Expert Meeting on Critical Limits for Heavy Metals and Methods for Their Application, Berlin, 2–4 December 2002. Held under the UN/ECE Convention on Long-Range Transboundary Air Pollution UBA-Texte 47/03. Umweltbundesamt, Berlin, 29–78.

De Vries W, Schütze G, Lofts S, Tipping E, Meili M, Römkens PFAM, Groenenberg JE (2005) Calculation of critical loads for cadmium, lead and mercury. Background document to a mapping manual on critical loads of cadmium, lead and mercury. Alterra Report 1104. Alterra, Wageningen, the Netherlands.

Di Toro DM, Allen HE, Bergman HL, Meyer JS, Paquin PR, Santore RC (2001) Biotic ligand model of the acute toxicity of metals. 1. Technical basis. Environ Toxicol Chem 20:2383–2396.

Dwane GC, Tipping E (1998) Testing a humic speciation model by titration of copper-amended natural waters. Environ Int 24:609–616.

Forbes TL, Forbes VE (1993) A critique of the use of distribution-based extrapolation models in ecotoxicology. Funct Ecol 7:249–254.

Giller KE, Witter E, McGrath SP (1998) Toxicity of heavy metals to microorganisms and microbial processes in agricultural soils: a review. Soil Biol Biochem 30:1389–1414.

Gregor HD, Spranger T, Hönerbach F (eds) (1997) Critical Limits and Effects-Based Approaches for Heavy Metals and Persistent Organic Pollutants. United Nations Economic Commission for Europe (UN/ECE), Convention on Long-Range Transboundary Air Pollution (CLRTAP), Task Force on Mapping (TFM). Proceedings of a European Workshop on Effects-Based Approaches for Heavy Metals, Bad Harzburg, Germany, 3–7 Nov. 1997. Federal Environmental Agency (Umweltbundesamt), Berlin, Germany, UBA-Texte 5/98.

Gregor H-D, Mohaupt-Jahr B, Hönerbach F (eds) (1999) Effects-Based Approaches for Heavy Metals. United Nations Economic Commission for Europe (UN/ECE), Convention on Long-Range Transboundary Air Pollution (CLRTAP), Task Force on Mapping (TFM), Schwerin, Germany, 12–15 Oct. 1999.

Groenenberg JE, Römkens PFAM, Tipping E, Pampoura T, de Vries W, Schütze G (2003) Transfer functions for the calculation of critical loads for lead, cadmium and mercury. In: Schütze G, Lorent U, Spranger T (eds) Proceedings, Expert Meeting on Critical Limits for Heavy Metals and Methods for Their Application, Berlin, 2–4 December 2002. Held under the UN/ECE Convention on Long-Range Transboundary Air Pollution UBA-Texte 47/03. Umweltbundesamt, Berlin, Germany, 79–102.

Hettelingh JP, Slootweg J, Posch M, Ilyin I (2002) Preliminary Modelling and Mapping of Critical Loads of Cadmium and Lead in Europe. RIVM Report 259101011. CCE MSC-East Moscow and CCE Bilthoven, National Institute of Public Health and Environmental Protection, Bilthoven, The Netherlands.

Jongbloed RH, Pijnenburg J, Mensink BJWG, Traas TP, Luttik R (1994) A model for environmental risk assessment and standard setting based on biomagnification.

Top predators in terrestrial ecosystems. Report 719101 012. National Institute of Public Health and the Environment, Bilthoven, The Netherlands.
Klepper O, van de Meent D (1997) Mapping the Potentially Affected Fraction (PAF) of Species as an Indicator of Generic Toxic Stress. RIVM Report 607504 001. National Institute of Public Health and the Environment, Bilthoven, the Netherlands.
Lanno RP, Conder JM, Seals L (1999) Critical limits for heavy metals in soils and surface waters. In: Proceedings of Workshop on Effects-Based Approaches for Heavy Metals. Schwerin, Germany, 12–15 October 1999, pp 49–56.
Lee JG, Roberts SB, Morel FMM (1995) Cadmium, a nutrient for the marine diatom *Thalassiosira weissflogii*. Limnol Oceanogr 40:1056–1063.
Lijzen JPA, Mesman M, Aldenberg T, Mulder CD, Otte PF, Posthumus R, Roex E, Swartjes FA, Versluijs CW, van Vlaardingen PLA, van Wezel AP, van Wijnen HJ (2002) Evaluatie onderbouwing BodemGebruiksWaarden. RIVM rapport 711701 029. Rijksinstituut voor Volksgezondheid en Milieuhygiëne, Bilthoven, Netherland.
Lofts S, Spurgeon DJ, Svendsen C, Tipping E (2004) Deriving soil critical limits for Cu, Zn, Cd and Pb: a method based on free ion concentrations. Environ Sci Technol 38:3623–3631.
Ma W, van der Voet H (1993) A risk-assessment model for toxic exposure of small mamalian carnivores to cadmium in contaminated natural environments. Sci Total Environ (Suppl) 1993:1701–1714.
Mance G (1987) Pollution Threat of Heavy Metals in Aquatic Environments. Elsevier Applied Science, London.
McLaughlin M, Hamon R, Lombi E, Smolders E (2004) Metal Ecotoxicity Protocols in Soil—Rubbing Salt into the Wounds. SETAC Globe, pp 51–52.
Meili M (1991) The coupling of mercury and organic matter in the biogeochemical cycle—towards a mechanistic model for the boreal forest zone. Water Air Soil Pollut 56:333–347.
Meili M (1997) Mercury in lakes and rivers. In: Sigel A, Sigel H (eds) Mercury and Its Effects on Environment and Biology. Marcel Dekker, New York, pp 21–51.
Meili M, Åkerblom S, Bringmark L, Johansson K, Munthe J (2003a) Critical loads and limits of heavy metals in ecosystems: some Swedish contributions to European modelling efforts. Background document contributed to the Editorial Meeting of the Expert Panel on Critical Loads of Heavy Metals under UN/ECE-CLRTAP-ICP Modelling and Mapping, Paris, 9–10 April 2003, http://www.oekodata.com/pub/mapping/workshops/ws_berlin/sweden.pdf.
Meili M, Bishop K, Bringmark L, Johansson K, Munthe J, Sverdrup H, de Vries W (2003b) Critical levels of atmospheric pollution: criteria and concepts for operational modelling of mercury in forest and lake ecosystems. Sci Total Environ 304:83–106.
Morel FMM (1983) Principles of Aquatic Chemistry. Wiley, New York.
Newman MC, Ownby DR, Mézin LCA, Powell DC, Christensen TRL, Lerberg SB, Anderson BA (2000) Applying species-sensitivity distributions in ecological risk assessment: Assumptions of distribution type and sufficient numbers of species. Environ Toxicol Chem 19:508–515.
OECD (Organisation for Economic Cooperation and Development) (1989) Report of the OECD Workshop on Ecological Effects Assessment. Environment Monographs No. 26. OECD, Paris.

OECD (Organisation for Economic Cooperation and Development) (1992) Report of the OECD Workshop on the Extrapolation of Laboratory Aquatic Toxicity Data on the Real Environment. Environment Monographs No. 59. OECD, Paris.

Palmborg C, Bringmark L, Bringmark E, Nordgren A (1998) Multivariate analysis of microbial activity and soil organic matter at a forest site subjected to low-level heavy metal pollution. Ambio 27:53–57.

Prasad MNV (1995) Cadmium toxicity and tolerance in vascular plants. Environ Exp Bot 35:525–545.

Ritchie GSP, Sposito G (2001) Speciation in soils. In: Ure AM, Davidson CM (eds) Chemical Speciation in the Environment. Blackwell, Oxford, pp 237–264.

Römkens PFAM, Groenenberg JE, Bonten LTC, de Vries W, Bril J (2004) Derivation of partition relationships to calculate Cd, Cu, Ni, Pb and Zn solubility and activity in soil solutions. Alterra rapport 305. Alterra, Wageningen.

Santore RC, Di Toro DM, Paquin PR, Allen HE, Meyer JS (2001) Biotic ligand model of the acute toxicity of metals. 2. Application to acute copper toxicity in freshwater fish and *Daphnia*. Environ Toxicol Chem 20:2397–2402.

Sauvé S, McBride MB, Hendershot WH (1997a) Speciation of lead in contaminated soils. Environ Pollut 98:149–155.

Sauvé S, McBride MB, Norvell WA, Hendershot WH (1997b) Copper solubility and speciation of in situ contaminated soils: effects of copper level, pH and organic matter. Water Air Soil Pollut 100:133–149.

Sauvé S, Norvell WA, McBride MB, Hendershot WH (2000) Speciation and complexation of cadmium in extracted soil solutions. Environ Sci Technol 34: 291–296.

Saxe JK, Impellitteri CA, Peijnenburg WJGM, Allen HE (2001) Novel model describing trace metal concentrations in the earthworm, *Eisenia andrei*. Environ Sci Technol 35:4522–4529.

Skyllberg U, Qian J, Frech W, Xia K, Bleam WF (2003) Distribution of mercury, methyl mercury and organic sulphur species in soil, soil solution and stream of a boreal forest catchment. Biogeochemistry 64:53–76.

Smolders E, Buekers J, Oliver I, McLaughlin MJ (2004) Soil properties affecting microbial toxicity of zinc to soil microbial properties in laboratory-spiked and field-contaminated soils. Environ Toxicol Chem 23:2633–2640.

Sola F, Isaia J, Masoni A (1995) Effects of copper on gill structure and transport function in the rainbow trout, *Oncorhynchus mykiss*. J Appl Toxicol 15: 391–398.

Spurgeon DJ, Hopkin SP (1996) Effects of variations of the organic matter content and pH of soils on the availability and toxicity of zinc to the earthworm *Eisenia fetida*. Pedobiology 40:80–96.

Stevens DP, McLaughlin MJ, Heinrich T (2003) Determining toxicity of lead and zinc runoff in soils: salinity effects on metal partitioning and on phytotoxicity. Environ Toxicol Chem 22:3017–3024.

Tambasco G, Sauvé S, Cook N, McBride M, Hendershot W (2000) Phytoavailability of Cu and Zn to lettuce (*Lactuca sativa*) in contaminated urban soils. Can J Soil Sci 80:309–317.

Thakali S, Allen HE, Di Toro DM, Ponizovsky A, Rooney C, Zhao F-J, McGrath SP, Criel P, van Eeckhout H, Janssen C, Oorts K, Smolders E (2005) Development of terrestrial biotic ligand models for copper and nickel toxicity in soils:

application for plant, invertebrate, and microbial tests. In: Abstracts, SETAC Europe 15th Annual Meeting, Lille, France, 22–26 May, p. 53.
Tipping E (1994) WHAM—a chemical equilibrium model and computer code for waters, sediments, and soils incorporating a discrete site/electrostatic model of ion-binding by humic substances. Comput Geosci 20:973–1023.
Tipping E (1998) Humic ion-binding Model VI: an improved description of the interactions of protons and metal ions with humic substances. Aquat Geochem 4:3–47.
Tipping E (2002) Cation Binding by Humic Substances. Cambridge University Press, Cambridge.
Tipping E (2005) Modelling Al competition for heavy metal binding by dissolved organic matter in soil and surface waters of acid and neutral pH. Geoderma 127:293–304.
Tipping E, Woof C, Hurley MA (1991) Humic substances in acid surface waters; modelling aluminium binding, contribution to ionic charge-balance, and control of pH. Water Res 25:425–435.
Tipping E, Rey-Castro C, Bryan SE, Hamilton-Taylor J (2002) Al(III) and Fe(III) binding by humic substances in freshwaters, and implications for trace metal speciation. Geochim Cosmochim Acta 66:3211–3224.
Tipping E, Lofts S, Smith EJ, Shotbolt L, Ashmore MR, Spurgeon D, Svendsen C (2003) Information and proposed methodology for determining critical loads of cadmium and lead; a UK contribution. Background document contributed to the UNECE ICP Modelling and Mapping Meeting, Paris, April 9–10, 2003.
Tyler G (1992) Critical concentrations of heavy metals in the mor horizon of Swedish forests. Report 4078. Swedish Environmental Protection Agency, Solna, Sweden.
Van Gestel CAM, Koolhaas JE (2004) Water-extractability, free ion activity, and pH explain cadmium sorption and toxicity to *Folsomia candida* (Collembola) in seven soil-pH combinations. Environ Toxicol Chem 23:1822–1833.
Vig K, Megharaj M, Sethunathan N, Naidu R (2003) Bioavailability and toxicity of cadmium to microorganisms and their activities in soil: a review. Adv Environ Res 8:121–135.
Vulkan R, Zhao F-J, Barbosa-Jefferson V, Preston S, Paton GI, Tipping E, McGrath SP (2000) Copper speciation and impacts on bacterial biosensors in the pore water of copper-contaminated soils. Environ Sci Technol 34:5115–5121.
Weng LP, Temminghoff EJM, van Riemsdijk WH (2001) Contribution of individual sorbents to the control of heavy metal activity in sandy soil. Environ Sci Technol 35:4436–4443.
Weng LP, Temminghoff EJM, Lofts S, Tipping E, van Riemsdijk WH (2002) Complexation with dissolved organic matter and solubility control of metals in a sandy soil. Environ Sci Technol 36:4804–4810.
Witter E (1992) Heavy metal concentrations in agricultural soils critical to microorganisms. Report 4079. Swedish Environmental Protection Agency, Solna, Sweden.

Manuscript received July 21; accepted August 2, 2006.

Rev Environ Contam Toxicol 191:91–130

Critical Soil Concentrations of Cadmium, Lead, and Mercury in View of Health Effects on Humans and Animals

Wim de Vries, Paul F.A.M. Römkens, and Gudrun Schütze

Contents

Communicated by Pim de Voogt.

W. de Vries (✉), P.F.A.M Römkens
Alterra, Wageningen University and Research Centre, Droevendaalse steeg 4, Atlas 104, P.O. Box 47, NL-6700 AA Wageningen, The Netherlands

G. Schütze
ÖKO-DATA Ecosystem Analysis and Environmental Data Management, Hegermühlenstraße 58, 15344 Strausberg, Germany

I. Introduction

To assess the impact of elevated concentrations of metals in terrestrial ecosystems, a major distinction should be made in risks/effects of heavy metals related to (i) the soil ecosystem (soil organisms/processes and plants) and (ii) human health or animal health resulting from bioaccumulation. The latter effect is related to the phenomenon that a chemical accumulates in species through different trophic levels in a food chain, or secondary poisoning. Heavy metal accumulation in the food chain is specifically considered important with respect to cadmium (Cd), mercury (Hg), and, to a lesser extent, lead (Pb). Accumulation ultimately causes toxic effects on (i) humans by affecting food quality of crops (Kawada and Suzuki 1998) and animal products, as well as drinking water quality, and (ii) animal health by affecting fodder quality and by direct intake of contaminated soil (Adriano 2001). For both humans and animals, health effects arise mainly through accumulation in target organs such as kidney and liver (Satarug et al. 2000). Apart from direct health effects related to intake of food and soil, elevated metal levels in soil also lead to an increase in leaching losses of metals to groundwater and surface water, which will, after a considerable delay time, affect both drinking water quality and aquatic organisms (Crommentuijn et al. 1997).

An overview of the pathways of metals in terrestrial and aquatic ecosystems, including the pathways, considered in this review, is given in Table 1

Table 1. Receptors of Concern in Three Main Types of Terrestrial Ecosystems.

Receptors of concern	Type of ecosystem: Arable land	Grassland	Nonagricultural land
Ecosystem			
Soil microorganisms	+	+	+
Soil invertebrates	+	+	+
Plants	+	+	−
Wild plants	−	−	+
Human health/animal health			
Plants			
Food crops (human health)	+	−	−
Fodder crops (animal health)	−	+	−
Groundwater[a] (human health)	+	+	+
Animals			
Cattle (human animal health)	−	+	+
Birds/mammals (animal health)	+	+	+

[a]Refers specifically to groundwater used as drinking water.

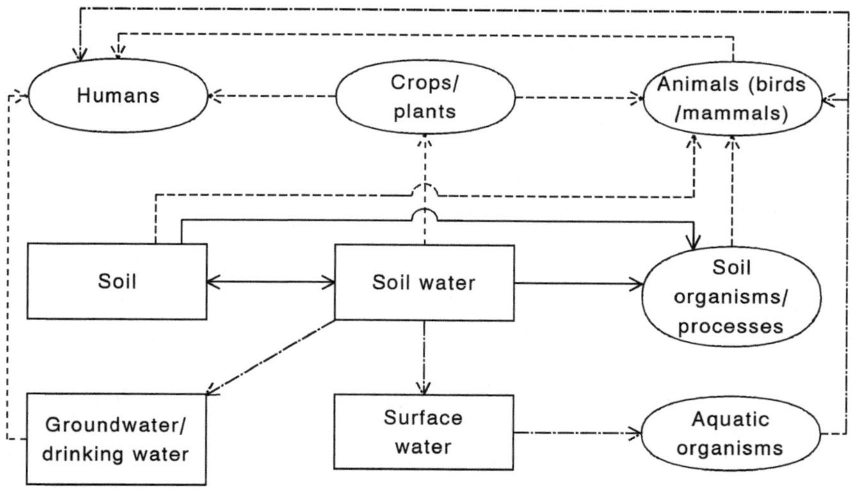

Fig. 1. Overview of the relationships and fluxes of metals from the soil to other compartments in terrestrial and aquatic ecosystems. Solid arrows are related to impacts on the soil ecosystem, discussed in another chapter (De Vries et al., this volume), whereas the dotted arrows refer to impacts on the health (or food quality) of plants, animals, and humans caused by accumulation in the food chain, discussed in this review. The hatched arrows are not considered as the paper focuses on terrestrial ecosystems.

and Fig. 1. This Chapter is limited to interactions in the soil (terrestrial ecosystems as well as agricultural systems) only. Hence, risks/effects of heavy metal uptake by consumption of aquatic organisms such as fish, which is specifically relevant for Hg (Meili 1997; Meili et al. 2003), is mentioned but no relationship can be established with the metal content in soil. Uptake of metals by humans is not included in Fig.1, although it plays a role for young children, especially for lead. Direct intake of soil in, for example, home gardens, is considered an important factor. The current Dutch soil quality standards for lead for home's gardens are in fact based partially on the ingestion of soil by children (VROM 1998). This aspect, however, is beyond the scope of this review. Here, we focus on the relationships in the soil–plant–animal chain, and exposure of humans to metals is considered within this "system" only. The relevant receptors in terrestrial ecosystems are presented in Table 1, with receptors considered here shown in bold.

At present, the transfer of metals, and chemicals in general, in food chains is generally described by bioconcentration factors (BCFs) and bioaccumulation factors (BAFs). These factors are based on the assumption of the existence of linear relationships between metals in soil and those in plants, earthworms, mammals, and other target organisms. The BCF is generally defined as the ratio of the test chemical concentration in an organism (e.g., plant, earthworm) to the concentration in water or soil at steady state. The BAF is defined as the ratio of the test chemical concentration in an organism to the concentration in its food at steady state (Jongbloed et al. 1994). BCFs are generally used for plants and invertebrates and are expressed in wet weight of tissue and dry weight of soil, whereas BAFs are generally used for accumulation by birds and mammals and are expressed on a wet weight basis.

The aim of this review is to illustrate that use of such constant accumulation factors is generally not adequate to describe metal transfer in the food chain. Here we show that there is a need to account for differences in soil properties when describing relationships between metal concentrations in soil and those in plants, water, and soil organisms. More specifically, we show how critical concentrations for Cd, Pb, and Hg in soil, in view of their potential impacts on human health and on animal health, can be derived by accounting for differences in soil properties. An analogous approach has been described by de Vries et al. (this volume) to derive critical concentrations for metals in soil and soil solution in view of ecotoxicological impacts on soil organisms and plants. Section II focuses on the derivation of critical concentrations for metals in soil in view of human health effects, resulting from intake of food crops, animal products, and drinking water. Section III is dedicated to the derivation of critical soil metal concentrations related to impacts on animal health, focusing on simple food chain models for birds and mammals feeding on worms and/or plants. In deriving such critical soil metal concentrations, use has been made of quality criteria or target values in crops and terrestrial fauna, which have been back-calculated to the soil using soil–plant, soil–soil invertebrate, plant–animal, and soil invertebrate–animal relationships, as discussed in detail in those subsections. We end with a critical evaluation of the assumptions related to the derivation and use of the critical soil metal concentrations (Section IV).

II. Critical Soil Metal Concentrations Related to Impacts on Human Health

Critical soil metal concentrations related to human toxicological effects can be derived from critical limits for humans (e.g., acceptable daily intake, ADI, in μg/kg/d) with an integrated model in which all relevant exposure pathways have been included. The ADI is the quantity of a compound to which man can be orally exposed, on the basis of body weight, without experiencing adverse effects on health. An example of such a model is CSOIL (Van den Berg and Roels 1991; Rikken et al. 2001), which derives a critical limit for soil from a given ADI value. This model includes many exposure routes to humans, such as intake from crops, meat, drinking water, and air and soil ingestion. The derivation of a critical soil limit related to an ADI by the CSOIL model depends strongly on many assumptions regarding the intake of food (Lijzen et al. 2001).

Here, it is assumed that critical soil metal concentrations related to human toxicological effects can be derived adequately from back-calculating food quality criteria (mg/kg) for metals in food crops, animal products (meat or milk), or target organs (e.g., kidney and liver) of cows/sheep and drinking water quality criteria (μg/L). Food quality and drinking water criteria quality criteria are thus used as an alternative to ADIs for humans to derive critical soil metal concentrations, thus avoiding the need of a comprehensive model on human exposure pathways. Next, we present an overview of quality criteria including a description how food quality criteria for crops can be derived from ADIs (II.A). We then illustrate how critical metal (Cd, Pb, and Hg) concentrations in soil can be derived as a function of soil properties from (i) food quality of crops (II.B), (ii) food quality of animal products, with an emphasis on grazing cows and sheep (II.C), and (iii) drinking water quality (II.D).

A. Health Impacts and Quality Criteria

For metals such as Cd, Pb, and Hg, no biological function is known. The possible health effects of exposure to cadmium, lead, and mercury have been investigated for many years, both for humans and for animals. A summary of those effects, based on studies summarized in reports published by, for example, the World Health Organization, the International Agency for Research on Cancer, the U.S Department of Health and Human Services, and Centers for Disease Control, is given in Jakubowski (2003). The major routes for human exposure are consumption of food, drinking of water, and, to a lesser extent, inhalation of air and intake of soil (children). In general, food is the dominant route of exposure of Cd and Pb of non-smokers (tobacco smoking can at least double the Cd intake), whereas the intake of fish is an important route of Hg intake (Anonymous 2000). In case of Cd and inorganic Hg, the kidney is the most sensitive and therefore most important target organism to protect. The target site for Pb toxicity is cognitive impairment associated with Pb levels in blood above 100 μg/L. This section contains an overview and discussion on (i) existing concepts that can be used to derive food quality criteria from ADI values and (ii) critical limits for Cd, Pb, and Hg in food crops and drinking water.

Derivation of Food Quality Criteria from Acceptable Daily Intakes

Food quality criteria, combined with soil–plant relationships, can be used to derive critical soil limits, thus avoiding the use of a detailed human exposure model while still using the concept of ADI. Next, we illustrate the relationship between food quality criteria and ADIs with the example of Cd in wheat. Wheat is considered one of the most important exposure pathways for humans.

Table 2. Relationships between Food Quality Criteria for Grain and Fish and Acceptable Daily Intakes (ADI).

Symptom incidence percentage	Cd limits		
	ADI_{total} (μg/d)	ADI_{grain} (μg/d)	Grain content[a] (mg/kg)
0.2	40	20	0.33
0.05	28	14	0.23
0.02	15	7.5	0.12
0.01	9	4.5	0.08

[a]Based on dividing the ADI by a net grain intake of 60 g/d (400 g/d times a body uptake efficiency of 15%).

Dose–response data for symptom incidence by humans for Cd exposure are presented by Sverdrup (2002). For Cd, the symptom is expressed as percent (%) incidence of tubular proteuria at the age of 40 yr. For Cd, ADIs are given as microgram per day (μg/d) for adult persons. Table 2 shows how the ADI values, related to incidence levels, that vary between 0.01% and 0.2% for Cd can be transferred to food quality criteria for grain, depending on the percent incidence accepted.

To perform the calculations for Cd, it is assumed that the total diffuse background exposure resulting from other exposure pathways is approximately equal to the exposure caused by eating bread or fish. For Cd, another important pathway is drinking water. In performing the calculations to derive a critical Cd content in grain, the daily intake of grain is set at 400 g and the body uptake efficiency, defined as the ratio between the total amount taken up by the body and the total administered dose, is assumed to equal 15%. The uncertainty of this coefficient, however, is large, and it can vary from 5% to 20% (Friberg et al. 1979). To derive a critical Hg content in fish, a weekly intake of 200 g of fish is assumed.

Based on a symptom incidence of 0.01%, being the standard risk level accepted for generic medical preparates, the acceptable Cd content in grain equals 0.08 mg/kg. Using a 10-fold-higher acceptable risk level (0.1%), the content increases to 0.28 mg/kg. The recommended food quality criterion for Cd in grain is 0.20 mg/kg (formerly 0.10 mg/kg), Which shows that the range in food quality criteria is in line with the range of calculated values.

Critical Limits for Cadmium, Lead, and Mercury in Food and Drinking Water

The major routes for human exposure are consumption of food, drinking of water, and inhalation of air, in case of smoking. The latter aspect is not considered here. In Table 3, an overview is given of relevant critical limits

Table 3. Overview of Food, Drinking Water, and Air Quality Criteria for Cd, Pb, and Hg in View of Human Health Effects.

		Critical limit			
Receptor	Unit[1]	Cd[a]	Pb[a]	Hg[a]	Source
Wheat	mg/kg	0.20 (0.10)[b]	0.2	0.03	Food quality criteria, EU 2001
Vegetables[c]	mg/kg	0.20	0.3	0.03	Food quality criteria, EU 2001
Drinking water	μg/L	3	10	1	WHO 2004

[a]All critical limits for food and fish are in mg/kg fresh weight.

[b]For wheat, the current critical limit is 0.20 mg/kg but we also investigated resulting critical soil concentrations using previous value of 0.10 mg/kg.

[c]Examples are endive, spinach, lettuce, etc.

for Cd, Pb, and Hg in this context, focusing on wheat in the case of food. Food is the main source of cadmium exposure in the general population (about 94%–99% of the total intake in nonsmokers). In this context, wheat is an important food product and, because wheat tends to accumulate Cd rather easily in the grain, this leads to the most sensitive critical limit for soil.

The EU regulation (EG) No. 466/2001 uses a limit for Cd of 0.2 mg/kg fresh weight in wheat grains. This limit was based on the principle "As Low As Reasonably Achievable" (ALARA) and is, therefore, not based on effects. There are, however, indications that from the point of view of protection of human health, the critical limit of 0.1 mg/kg fresh weight, which was used in the EU before 2001, is more appropriate. In the European Draft Risk Assessment Report for Cd Metal and Cd-Oxides (RAR-Cd; EC 2003), a Cd ADI for adult nonsmokers of 37–47 μg, referring to a body weight of 55–70 kg, respectively, was used to assess the current and future risk of Cd to populations by environmental exposure on a regional and continental scale. Assuming that 50% of the Cd ADI intake can be filled by grain diets, and assuming a daily intake of 200 g grains, the critical Cd content in grain would be 0.09–0.12 mg/kg wet wt. The assumed daily consumption of 200 g cereals is based on a range of 142–266 g in different European countries (EC 2003).

The assumption of 50% dietary Cd intake by wheat is based on (i) an estimate that 30%–50% of the dietary intake of Cd of the German population stems from consumption of flour and its products (Schütze et al. 2003a) and (ii) the fact that the calculated daily dietary Cd intake of 16–22 μg/d for Germans was in good agreement with the values reported in the RAR-Cd for the European population, which ranged between 10 and 21 μg/d. This value could be derived from German studies on food consumption behavior

(Kübler et al. 1995; Statistisches Bundesamt 1999) and data from the German food monitoring program (BgVV 1997). Consequently, an effects-based critical limit for Cd in wheat of 0.1 mg/kg wet wt was recommended for use in the framework of the Convention on Long-range Transboundary Air Pollution (Schütze et al. 2003b). In this study, we investigated the impact of using food quality criteria of both 0.20 mg/kg and 0.10 mg/kg fresh weight wet wt in wheat on critical soil Cd concentrations. A discussion on the reliability of the limits is continued in Section IV.

B. Derivation of Critical Soil Metal Concentrations from Food Quality Criteria for Crops

Approach

Figure 2 contains a schematic representation of the link between critical metal concentrations for soil and those for crops. A distinction was made between (i) food quality criteria in view of human health, (ii) fodder quality criteria in view of animal health, and (iii) phytotoxic levels in view of toxic effects on the crop itself. The latter aspect is not related to human health

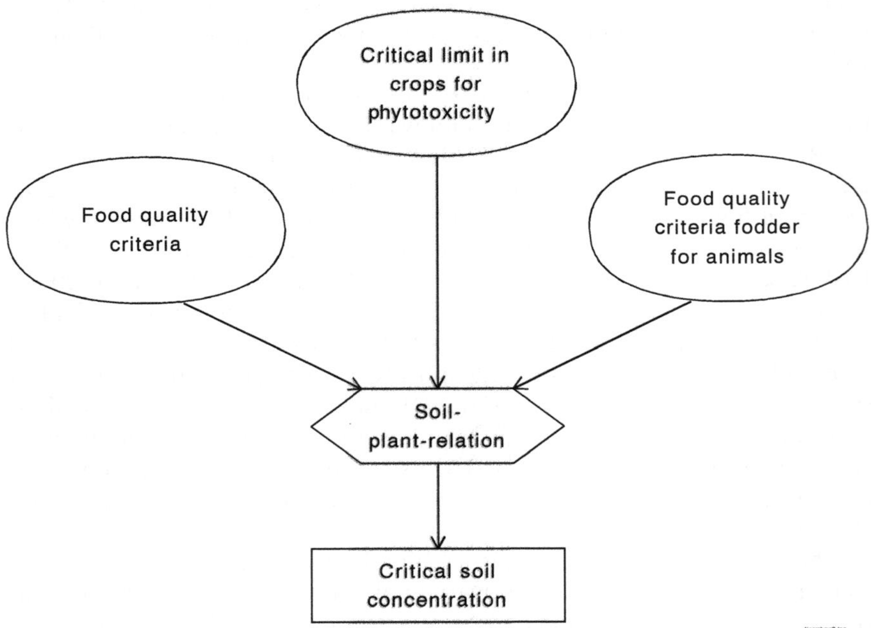

Fig. 2. Procedure that has been applied to derive critical limits for heavy metals in the soil from quality criteria in food crops in view of effects on humans (arable land) and in fodder in view of effects on animals (arable and grassland) and from critical limits in crops in view of phytotoxic effects (grassland and arable land).

but was included to be sure that the food quality criteria do not lead to situations where food crops are adversely affected. Because metal uptake by crops is plant specific, the kind of crop influences the derived limit for soil. It is thus necessary to derive relationships for the most sensitive crops to assess critical soil metal concentrations.

In most bioaccumulation models, including the aforementioned CSOIL model, a simple bioconcentration factor (BCF), often denoted as bioaccumulation factor (BAF), is used to calculate a metal content in plants from a total metal concentration in soil according to the following equation:

$$[M]_p = BCF_{sp} \cdot [M]_{s,tot} \quad (1)$$

where:

$[M]_{p(crit)}$ = metal concentration in plant (mg/kg)
$[M]_{s(crit)}$ = total metal concentration in soil (mg/kg)
BCF_{sp} = bioconcentration factor from soil to plant, being the ratio of metal concentration in plant to total metal concentration in soil (–)

Often a median BCF value based on many plant and soil data is used. Such an approach is only acceptable if a linear relationship between plant and soil content has been proven to exist, based on data. However, field data from various studies have shown that for most metals such a linear relationship does not exist. For certain metal–plant combinations, there is no relationship between soil and plant at all. To illustrate the absence of a simple relationship between metal contents in plant and soil, Fig. 3 gives an overview of Cd, Pb, and Hg contents in grass and wheat and in soil from Dutch agricultural fields. A poor relationship can be discerned only For Cd, but for Pb and Hg the BCF approach does not work. Instead, other factors, including above-ground uptake metals from atmospheric deposition, may be more important. For Pb, direct uptake is specifically relevant for vegetables, and for Hg it is often assumed that crop uptake is completely controlled by atmospheric deposition (De Temmerman and de Witte 2003a,b). Such relationships can be used to derive critical limits for these metals in the air, as summarized in De Vries et al. (2005). Furthermore, some crops can actively reduce the availability of metals in the rooting zone, thus reducing the application of any soil to plant relationship. When a relationship between the plant and soil heavy metal content is absent, it is impossible to derive critical soil metal concentrations from critical limits in plants and the use of a BCF gives a false impression of a limit thus derived.

Apart from the erroneous use of BCF values when a relationship is absent, it is often also inadequate to use BCFs even when such a relationship exists. In that case, better predictions of the metal content in plants can be obtained by a nonlinear relationship accounting for the impact of important soil properties that control the bioavailability of metals in soils, according to Brus et al. (2002) and Adams et al. (2004):

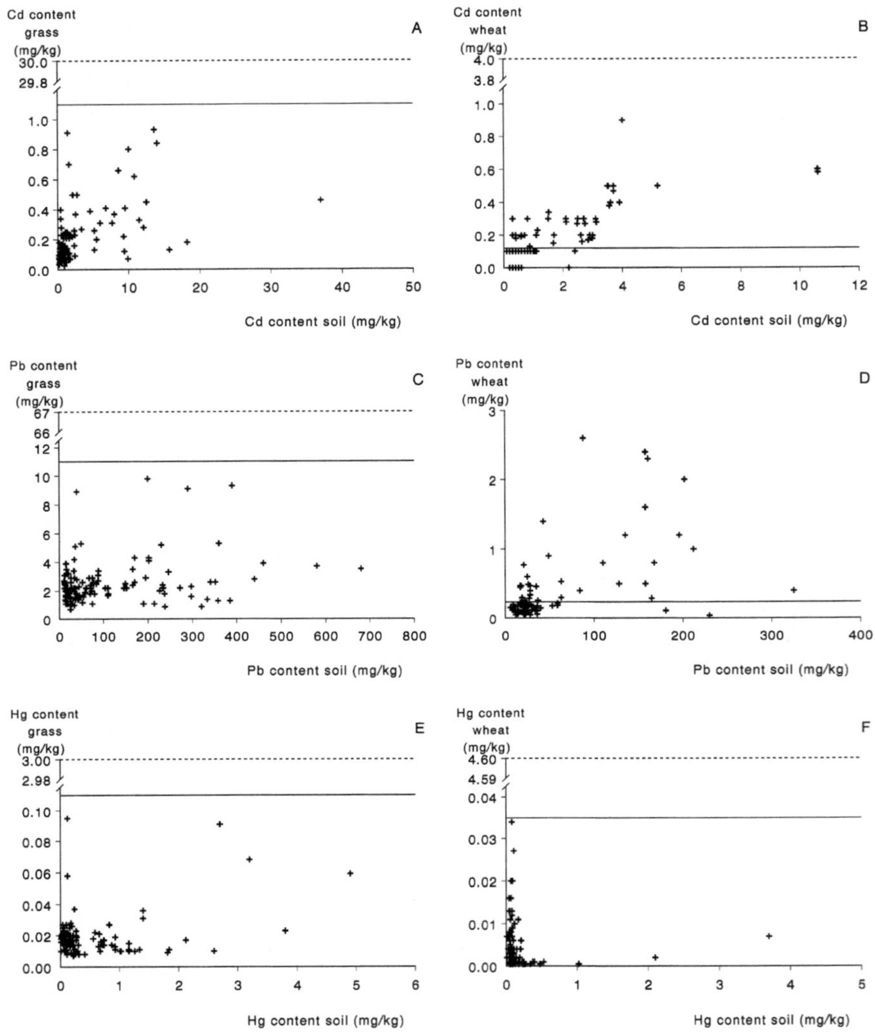

Fig. 3. Relationships between Cd, Pb, and Hg contents in grass and soil (A, C, E) and in wheat and soil (B, D, F). The solid line represents the fodder or food quality criteria as in Table 4, and the dashed line represents limits in view of phytotoxic effects on crops.

$$[M]_p = K_{sp} \cdot [M]_{s,tot}^{n} \quad (2)$$

where:

K_{sp} = transfer constant from soil to plant ($mg.kg^{1-n}$)
n = coefficient describing the nonlinear relationship (–)

in which the value of K_{sp} depends on the content of organic matter and clay and the soil pH, according to:

$$\text{Log } K_{sp} = a + b \cdot pH - KCl + c \cdot \log[\text{clay}] + d \cdot \log[\text{OM}] \quad (3)$$

where:

[OM] = organic matter content in the soil (%)
[clay] = clay content in the soil (%)

A critical soil metal concentration can thus be calculated from the inverse nonlinear soil–plant relationship using a critical limit in a crop (food quality criteria, fodder criteria, or phytotoxicity limit), according to the following:

$$[M]_{s,tot(crit)} = ([M]_{p(crit)} / K_{sp})^{1/n} \quad (4)$$

where

$[M]_{p(crit)}$ = critical metal concentration in plant (mg/kg)
$[M]_{s(crit)}$ = critical total metal concentration in soil (mg/kg)

For each combination of crop and heavy metal, an evaluation of the resulting regression equation is needed to decide whether the predictions are accurate enough to be used in this approach. Furthermore, application of an inverse regression equation is warranted only when maximum measured metal contents in plants, used in deriving the relationship, do approach (and preferably exceed) the critical limits in plants. Otherwise, the derivation of critical soil metal concentrations from critical plant contents implies that the relationship is applied outside its range of derivation, which may lead to highly unreliable results (De Vries et al. 2007).

For Cd, Pb, and Hg in grass, maize, sugar beet, wheat, potatoes, lettuce, endive, and spinach, these being the main crops in the Netherlands, relationships were derived with the total soil concentration according to the following (see also Eqs. 2 and 3; Brus et al. 2002; Adams et al. 2004):

$$\log[M]_p = a + b \cdot pH + c \cdot \log[\text{clay}] + d \cdot \log[\text{OM}] + n \cdot \log[M]_{s,tot} \quad (5)$$

Values for the various coefficients (the exponent n and the parameters a, b, c, and d) were derived by multiple regression analyses. To derive these equations, data from field studies were used (Wiersma et al. 1986; Van Driel et al. 1988). In contrast to many laboratory studies, the soil samples from these field sites were not amended with metals and therefore reflect "real" field conditions. More information on the approach and datasets is given in De Vries et al. (2006). In general, relationships were reasonable to good for Cd, relatively poor for Pb, and absent for Hg. As an example, results for Cd and Pb for grass, maize, wheat, and lettuce are presented in Table 4.

For grass and maize, no relationships were found for Pb, and for wheat and lettuce the relations were very weak, most likely the result of processes

Table 4. Overview of Selected Soil–Plant Relationships for Cd and Pb.

Crop	Soil–plant relationship[a]	R^2
Grass		
Cd	$\log(Cd_{plant}) = 0.17 - 0.12{*}pH - 0.28{*}\log(OM) + 0.49{*}\log(Cd_{soil})$	0.53
Pb	No relationship found	—
Maize		
Cd	$\log(Cd_{plant}) = 0.9 - 0.21{*}pH - 0.32{*}\log(clay) + 1.08{*}\log(Cd_{soil})$	0.62
Pb	No relationship found	—
Wheat		
Cd	$\log(Cd_{plant}) = 0.35 - 0.15{*}pH - 0.39{*}\log(OM) + 0.76{*}\log(Cd_{soil})$	0.72
Pb	$\log(Pb_{plant}) = -0.25{*}pH - 1.42{*}\log(OM) + 1.14{*}\log(Pb_{soil})$	0.24
Lettuce		
Cd	$\log(Cd_p) = 2.55 - 0.33{*}pH - 0.19{*}\log(clay) - 0.39{*}\log(OM) + 0.85{*}\log(Cd_{soil})$	0.71
Pb	$\log(Pb_p) = -0.65 + 0.59{*}pH - 0.30{*}\log(OM) + 0.59{*}\log(Pb_{soil})$	0.40

[a]pH is pH_{KCl}, clay is clay content in %, and OM is organic matter content in %.

at the soil–root interface where lead uptake is actively blocked by plants. In situations with significant relationships, the sign of the coefficients (pH-KCl, clay, and OM) is negative, which implies that an increase in pH, clay content, and organic matter content leads to a lower metal content in crops. This result is in agreement with the impact of the aforementioned soil properties on the availability of metals in soil. For metals such as Cd and Pb, the availability and uptake by crops decreases with an increase in pH and organic matter or clay content.

Example of a Model Application

Here we illustrate how quality criteria in crops can be back-calculated to critical soil metal concentrations that are a function of soil properties (organic matter content, clay, content and soil pH), as these properties influence soil–plant relationships. An overview of the quality criteria (in mg/kg dry wt) used for the considered food crops (wheat, potato, lettuce, and endive) and fodder crops (grass, maize, and sugar beet) is given in Table 5. Table 5 also contains background information on the original food quality criteria given as fresh weight and an overview of critical limits in view of phytotoxic effects on crops, based on literature information. As expected, food and fodder quality criteria are much more stringent than limits in view of phytotoxic effects on crops. In De Vries et al. (2006), more detail information is given on the background of all the criteria.

As an example of the applicability of the methodology, critical Cd concentrations in soil have been calculated using the food quality criterion for lettuce and the relevant soil–plant relationship presented in Table 6. The

Table 5. Overview of Fodder and Food Quality Criteria for Cd, Pb, and Hg in View of Animal Health and Human Health and Limits in View of Phytotoxic Effects on Crops (all Limits are Given on the Basis of Dry Weight).

		Quality criteria (mg/kg dry weight)					
		Food/fodder			Phytotoxicity		
Land use	Crop	Cd[a]	Pb[a]	Hg[a]	Cd[b]	Pb[b]	Hg[b]
Grassland	Grass	1.1	11	0.11	30[f]	67[j]	3[j]
Arable land	Maize	1.1	11	0.11	25[f]	38[j]	0.6[j]
Fodder crops	Sugarbeet	1.1	11	0.11	5[c]	—	1[c]
Arable land	Wheat	0.24	0.24	0.035	4[f]	—	4.6[j]
Food crops		(0.12)					
	Potato	0.42	0.42	0.13	5[c]	13[j]	1[c]
	Lettuce	4.0	6.0	0.60	10[e]	140[e,j]	1[c]
	Endive	3.3	5.0	0.50	15[f]	17[j]	1[c]

[a]The fodder quality criteria of Cd, Pb, and Hg for grass, maize, and sugarbeet are originally given as 1, 10, and 0.1 on the basis of 12% moisture content (food quality criteria; EU 2001). These data have been back-calculated to dry weight. The food quality criteria for wheat, potato, lettuce, and endive are originally given as fresh weight. In back-calculating to dry weight, the following moisture percentages were applied: wheat, 85% for the grain (the edible part); potato, 24%; lettuce, 5%; endive, 6%. For Hg, the food quality criteria are not considered applicable recently.

[b]For all crops, values are lower limits of ranges in phytotoxic contents.

The limits are based on the following sources:

[c]Kabata-Pendias and Pendias (1992), general crop-unspecific overview.

[d]Mortvedt et al. (1991).

[e]Smilde (1976).

[f]MacNicol and Beckett (1985), content at 10% reduction in yield.

[g]Dijkshoorn et al. (1979), content at 10% reduction in yield.

[h]Chang et al. (1992), content at 50% reduction in yield.

[i]Sheppard (1992), content at different percentages reduction in yield.

[j]Sauerbek (1983), content at different percentages reduction in yield.

Table 6. Calculated Critical Cd Contents in Soil in View of the Food Quality Criterion for Lettuce as a Function of Soil Properties.

Clay content (%)	Organic matter content (%)	Critical Cd content in soil (in mg/kg) pH 5	pH 6	pH 7
2	2	0.61	1.4	3.3
2	5	0.88	2.1	4.8
2	10	1.2	2.7	6.4
20	2	1.9	4.4	10
20	5	2.8	6.5	15
20	10	3.7	8.6	20

example refers to a sandy soil with 2% clay and a clay soil with 20% clay. To illustrate the effect of differences in soil properties, the effect of low and high organic matter content (2%, 5%, and 10%) and pH (5, 6, and 7) was also established. Results suggest that it is essential to make a distinction in soil types considering their difference in soil properties. In acid sandy soils, the critical Cd content is below 1 mg/kg (Table 6), approaching the critical Cd content related to ecotoxicological impacts (see De Vries et al., this volume).

To illustrate impacts of major soil types, critical soil metal concentrations have been calculated on the basis of food quality criteria for Cd for the following three major soil types in agriculture:

Sandy soils (3% OM; 3% clay, pH_{KCl} 5.5)
Clay soils (3% OM; 25% clay, pH_{KCl} 6.5)
Peat soils (30% OM; 15% clay, pH_{KCl} 6.0)

Because of a rather strict food quality criterion for Cd in wheat and the fact that Cd accumulates rather easily in wheat grains, critical limits for wheat in soil are most strict with respect to food crops (Table 7). Using the present food quality criterion of 0.2 mg/kg fresh weight, results for sugar beet appear to just as strict. Using the previously used food quality criterion of 0.1 mg/kg fresh weight, wheat is always the most sensitive crop in terms of calculating critical soil metal concentrations. Wheat is a crop that is widely cultivated over Europe and has a relevant share in the total human food intake. Also, sufficiently adequate soil–plant relationships, compared to other edible parts of agricultural crops, exist that allow for the calculation of the critical soil metal content ($R^2 > 0.7$; see Table 4). Cd in wheat is,

Table 7. Calculated Critical Cd Contents in Soil in View of the Food Quality Criteria for Different Crops (as in Table 4) and Soil Types.

		Critical Cd content in soil (mg/kg)			
Land use	Crop	Sand	Clay	Peat	All soils
Grassland	Grass	9.3	37	14	37
Arable land	Maize	2.6	7.6	5.3	6.1
	Sugar beet	0.94	3.3	2.0	2.2
	Wheat[a]	1.1 (0.46)	1.8 (0.72)	4.6 (1.9)	2.6 (1.1)
	Potato	5.3	9.3	14	10
	Lettuce	1.5	5.8	9.5	6.4
	Endive	0.93	5.3	8.3	5.8

[a]Values in brackets for wheat are results of calculations with the previously used food quality criterion of 0.1 mg/kg fresh weight, whereas the standard results are calculated with the present value of 0.2 mg/kg fresh weight.

therefore, an appropriate indicator of human health effects of Cd on arable land. Phytotoxic concentrations of Pb and Cd in food crops are in all cases much higher than limits related to human health; thus, there is no need to investigate critical loads of metals related to phytotoxic effects.

C. Derivation of Critical Soil Metal Concentrations from Food Quality Criteria for Animal Products

Approach

Figure 4 shows how critical metal concentrations for the soil have been derived from food quality criteria in animal products/organs related to human health and from acceptable daily intake by animals related to animal health. The latter aspect was included to be sure that the food quality criteria for humans do not lead to situations in which animal health is adversely affected. The derivation was limited to grazing animals, which are most sensitive due to ingestion of soil in addition to intake of grass. Figure 4 shows that such a derivation requires information on quality criteria or target values for metals in animal products, grass and soil intake, and

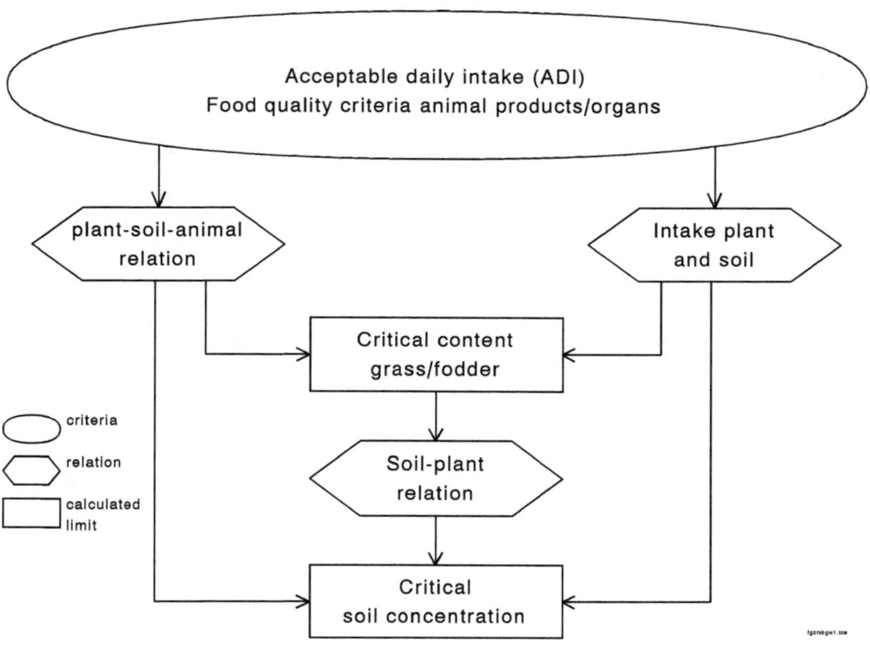

Fig. 4. Procedure that has been applied to derive critical limits for heavy metals in the soil (on grassland) from food quality criteria in animal products/organs in view of effects on humans and from acceptable daily intakes in view of toxic effects on animals.

soil–plant and plant–animal product relationships, including the effects of soil properties on these relationships.

When information is available on ADIs of metals, this can be used to obtain information on critical metal contents in fodder and soil according to:

$$[M]_{p(crit)} \cdot Ip + [M]_{s,tot(crit)} \cdot Is = ADI \qquad (6)$$

where:

ADI = acceptable daily intakes of metals (mg/d)
Ip = intake of plants (fodder) (kg/d)
Is = intake of soil (kg/d)

A combination of Eq. 2 and Eq. 6 gives:

$$K_{sp} \cdot [M]^{n}_{s,tot(crit)} \cdot Ip + [M]_{s,tot(crit)} \cdot Is = ADI \qquad (7)$$

From Eq. 7, the value of $[M]_{s,tot(crit)}$ can be solved iteratively on the basis of a given ADI and given values of K_{sp}, Ip, and Is. For those metals where no soil–plant relationship is available, median values of the metal content in the crop ($[M]_p$) can be used to calculate the corresponding critical soil metal content, according to:

$$[M]_{s,tot(crit)} = (ADI - [M]_p \cdot Ip) / Is \qquad (8)$$

When information is available on food quality criteria in animal organs/ products, this can be used to calculate an ADI by assuming the following:

- The availability of metals to animals is the same for metals present in plant products and soil, which implies that the transfer coefficients of metals from soil to animal product and that from plant to animal product is equal (see Fig. 4); this allows the calculation of an average concentration of metal in fodder, based on a certain intake of grass and the inevitable additional ingestion of soil.
- There is a direct linear relationship between metal content in animal organs/products and metal content in fodder (use of a BAF from plant to animal organ, BAF_{pa}).
- The intake of metals by other sources (water and air) is negligible.

Using these assumptions, the relationship between metal content in animal organs/products and in soil can be approximated as:

$$[M]_{ao(crit)} = \left(\frac{[M]_{p(crit)} \cdot Ip + [M]_{s,tot(crit)} \cdot Is}{Ip + Is} \right) \cdot BAF_{pa} \qquad (9)$$

where:

$[M]_{ao(crit)}$ = food quality criteria for metal content in animal organ (mg/kg)
BAF_{pa} = bioaccumulation factor from plant to animal organ/product (mg/ kg fresh weight/mg/kg dry weight)

A combination of Eq. 6 and 9 gives:

$$ADI = [M]_{ao(crit)} \cdot (Ip + Is) / BCF_{pa} \quad (10)$$

This again allows the calculation of $[M]_{s,tot(crit)}$, either iteratively from Eq. 7 or directly from Eq. 8.

Example of a Model Application

An example of a model application is limited to cows and sheep. Quality criteria or target values for metals in kidney, liver, and meat of those animals were used and back-calculated to the soil using soil–plant and plant–animal relationsships. In Table 8, an overview is given of the critical contents of Cd, Pb, and Hg used in view of food safety (food quality criteria) and animal health. An estimate of the ADI based on these criteria is given in Table 9 using Eq. 10 and the plant–animal bioconcentration factors given in the same table. The dry mass intake of grass by cows and sheep was assumed to be equal to 16.9 and 2.5 kg/d, respectively, and 0.41 and 0.10 kg/d of soil, assuming that the animals are always in the field ("worst case scenario"). Data for cows are based on McKone and Ryan (1989) and those for sheep on Huinink (2000).

Because there are no reliable soil–plant relationships for grass for any of the metals involved, critical soil concentrations of Cd, Pb, and Hg were calculated from the ADI, according to Eq. 8. To achieve this, measured median values of the metal content in grass were used (Wiersma et al. 1986). As the median plant metal content is hardly affected by the metal content of the soil or the soil type, the calculated critical soil concentrations hardly

Table 8. Critical Contents of Cd, Pb, and Hg in Animal Products and Animal Organs of Cows and Sheep in View of Food Safety (Food Quality Criteria, EU 2001) and Animal Health (Puls 1988) (all Limits are Given on the Basis of Fresh Weight).

		Critical limit (mg/kg)					
		Food safety			Animal health		
Animal	Organ	Cd	Pb	Hg[a]	Cd	Pb	Hg
Cow	Kidney	1.0	0.5	0.05	5	3	1.4
	Liver	0.5	0.5	0.05	1.4	2	2
	Meat	0.05	0.1	0.05	0.02	—	—
Sheep	Kidney	1.0	0.5	0.05	4	5	1
	Liver	0.5	0.5	0.05	2	5	4
	Meat	0.05	0.1	0.05	—	0.1	—

[a]For Hg, the food quality criteria have recently been abandoned. For sheep, the food quality criteria have been assumed equal to those for cows.

Table 9. Plant–Animal Bioconcentration Factors and Calculated Acceptable Daily Intake (ADI) of Cd, Pb, and Hg in Cows and Sheep in View of Impacts on Food Safety and Animal Health.

		BAF_{pa}[a,b] (mg/kg fresh weight/mg/kg dry weight)			ADI food safety (mg/d)			ADI animal health (mg/d)		
Animal	Organ	Cd	Pb	Hg	Cd	Pb	Hg	Cd	Pb	Hg
Cow	Unspecific	—	—	—	—	—	—	63[c]	2380[c]	28[c]
	Kidney	2.99	0.086	0.638	5.8	101	1.4	29	604	38
	Liver	0.554	0.0404	0.158	16	214	5.5	44	857	219
	Meat	3.3. 10^{-3}	1.3. 10^{-3}	9.2. 10^{-4}	262	1332	941	105	—	—
	Minimum	—	—	—	5.8	101	1.4	29	604	28
Sheep	Kidney	2.08	—	0.468	1.25	—	0.28	5	—	5.6
	Liver	1.85	—	0.0572	0.70	—	2.3	2.8	—	182
	Meat	2.9. 10^{-3}	—	9.4. 10^{-4}	45	—	138	—	—	—
	Minimum	—	—	—	0.70	—	0.28	2.8	—	5.6

[a]Estimates for BAF_{pa} for cows are based on Van Hooft (1995).

[b]Estimates for BAF_{pa} for sheep are based on Beresford et al. (1999). The values used are the upper estimates of the ranges given in this publication.

[c]Direct estimates for the ADI, unrelated to the various animals, are based on NOEC data in mg/kg bw/d for the oral intake of food and soil by Ma et al. (2001). The values were multiplied by a body weight of 70kg to get an ADI in mg/d.

Table 10. Overview of Critical Metal Contents in Soil on Grassland in View of Food Safety (Effects on Kidney) and Animal Health Calculated on the Basis of a Median Metal Plant Content and a Median BCF Value.

		Critical metal contents (mg/kg)			
		Food quality (kidney)		Animal health	
Metal	Soil type	$[M]_p(50\%)$	BCF (50%)	$[M]_p(50\%)$	BCF (50%)
Cd	Sand	10	1.3	67	6.5
	Clay	5.3	2.3	62	12
	Peat	9.6	2.0	66	10
Pb	Sand	155	41	1382	245
	Clay	155	124	1382	743
	Peat	159	106	1386	634
Hg	Sand	2.6	0.25	68	5.1
	Clay	2.7	1.3	68	28
	peat	2.5	0.68	67	14

differ between soils. This point is illustrated in Table 10, presenting calculated critical soil concentrations of Cd, Pb, and Hg based on the ADI in view of target values for the kidney of cows (the most sensitive animal organ) and in view of impacts on their health. For illustrative purposes, Table 10 also shows the values that would result from using median BCF values. Use of such values suggest an impact of soil type, but this only occurs because the ratio of metal contents in plant and soil differs between soil types without any real relationship involved. Results show that the critical soil metal concentrations are generally much higher than those derived from ecotoxicological impacts, as presented by De Vries et al. (this volume). Results for Cd and Hg for sheep (for Pb data that are lacking to allow the calculation, see Table 9) are highly comparable (De Vries et al. 2006). At the critical soil concentration derived by median plant metal contents, soil ingestion is the dominant pathway leading to critical metal concentrations in the kidney (calculated contribution, 61%–99%; De Vries et al. 2006).

D. Derivation of Critical Soil Metal Concentrations from Drinking Water Quality Criteria

Approach

The critical total Cd, Pb, and Hg concentration in soil related to human health effects can also be based on quality criteria (critical limits) for drinking water (WHO 2004). The WHO guideline includes the following quality criteria for Cd, Pb and Hg in view of drinking water quality: Pb 10 mg/m^3,

Cd 3 mg/m^3, and Hg 1 mg/m^3 (see Table 3). In several countries, such as the Netherlands, it is required that those concentrations should thus not be exceeded in groundwater used as drinking water. Based on the concentration in the soil pore water as a first estimate of the concentration in groundwater, an estimate of the related critical metal concentration in soil can be made using transfer functions that relate (i) the total dissolved metal concentration to the reactive soil metal concentration and (ii) the reactive soil metal concentration to the total soil metal concentration. Such transfer functions do exist for Cd and Pb (Römkens et al. 2004) as well as for Cu and Zn (these metals are not considered here) but not for Hg.

The reactive soil metal concentration can be derived from the total dissolved metal concentration according to:

$$[M]_{s,re} = K_f \cdot [M]_{ss}^{n} \tag{11}$$

where:

$[M]_{ss}$ = concentration of heavy metal M in the soil solution (mmol/L)
$M_{s,re}$ = reactive concentration of heavy metal M in the soil, in this case, a 0.43 M HNO_3 extractable content (mol/kg)
K_f = Freundlich coefficient (1/g * [(L^n)/($mmol^n$)])

The value of K_f is calculated as a function of the content of organic matter, clay, and pH extract according to:

$$\log K_f = \alpha_0 + \alpha_1 \cdot \log[OM] + \alpha_2 \cdot \log[clay] + \alpha_3 \cdot pH\ H_2O \tag{12}$$

where:

$\alpha_0 \ldots \alpha_3$ = regression coefficients
pH H_2O = pH in extract or soil solution

Values for the various regression coefficients were derived from laboratory experiments where soil samples were equilibrated with different extracting solutions at a 1:2 soil solution ratio. The result is a database with approximately 1400 soil soil solution records representing almost all Dutch soil types (Römkens et al. 2004). The coefficients for Eq. 12 are shown in Table 11.

Table 11. Values for the Coefficients α_0, α_1, α_2, α_3, and n in the Relationships Relating Dissolved Total Concentrations and Reactive Soil Concentrations of Cd and Pb, According to Eq. 12 after Römkens et al. (2004).

Metal	α_0	α_1	α_2	α_3	n	R^2	se(Y)
Cd	–4.85	0.58	0.28	0.27	0.54	0.79	0.33
Pb	–2.96	0.83	0.02	0.25	0.68	0.57	0.55

Table 12. Values for the Coefficients β_0-β_3 in the Relationships (Eq. 13) Relating Reactive, (0.43 N HNO_3), and Pseudo-total (Aqua Regia) Soil Concentrations of Cd and Pb, Using a Dutch Dataset (Römkens et al. 2004). The Relationships Hold for both $[M]_{tot}$ and $[M]_{re}$ in mg/kg.

Metal	β_0	$\beta_1\,[M]_{re}$	β_2 [OM]	β_3 [clay]	R^2_{adj}	Se(Y)[a]
Cd	0.028	0.877	0.009	0.081	0.96	0.10
Pb	0.323	0.810	0.035	0.136	0.92	0.13

[a]The standard error of the *y*-estimate on a logarithmic basis.

In this equation, dissolved organic carbon (DOC) was not included although it was available in the database. For applications on a regional or even national scale, however, data on DOC are usually not available, which was the main reason to exclude DOC from the model. For Cd, the effect of including DOC was rather small and the quality of predicted Cd concentrations was not significantly affected by removing DOC from the list of soil properties included. For Pb, however, the quality of model predictions was somewhat less when DOC was omitted from the equation. This result is not surprising because almost all Pb present in the soil solution is bound to DOC (Römkens et al. 2004).

In many countries, the regulation regarding critical metal concentrations in soil is based on total, aqua regia extractable, metal concentrations (actually being pseudo-total as aqua regia does not dissolve all metals). In the model relating the dissolved metal concentration to the metals in the solid phase, only the reactive fraction was included. To correct for this, the total metal concentration has to be derived from the criteria for the total contents because the total metal content equals the reactive and the not-reactive fraction. The total aqua regia extractable metal is derived from the reactive metal concentration and the content of organic matter and clay according to the following:

$$\log [M]_{s,tot} = \beta_0 + \beta_1 \cdot \log [M]_{s,re} + \beta_2 \cdot \log [OM]_s + \beta_3 \cdot \log [clay] \quad (13)$$

with the parameters given in Table 12 (see also the chapter by De Vries et al. in this volume).

Example of a Model Application

As an example of the applicability of the methodology, critical Cd and Pb contents have been calculated using drinking water quality standards and the relevant transfer functions presented in Table 13. The example refers to the same sandy soil and clay soil used before in deriving critical Cd contents in view of food quality criteria. Results again suggest show that it is essential to consider differences in soil types to derive relevant critical limits. In both acid and near-neutral sandy soils, the critical Cd content is below 1 mg/kg

Table 13. Calculated Critical Total Cd and Pb Contents in Soil in View of Drinking Water Quality Criteria as a Function of Soil Properties.

Clay content (%)	Organic matter content (%)	Critical Cd content (in mg/kg)		Critical Pb content (in mg/kg)	
		pH 4	pH 6	pH 4	pH 6
2	2	0.17	0.52	8.4	21
2	5	0.28	0.83	16	41
2	10	0.40	1.2	26	66
20	2	0.37	1.1	12	30
20	5	0.59	1.8	23	58
20	10	0.85	2.5	37	94

Table 14. Calculated Critical Total Cd and Pb Contents in Soil in View of Drinking Water Quality Criteria for Different Soil Types.

Metal	Soil use	Critical metal content (mg/kg)		
		Sand	Clay	Peat
Cd	Agriculture	0.55	1.9	4.0
	Nature	0.24	1.2	1.4
Pb	Agriculture	24	53	196
	Nature	12	31	77

(Table 13), approaching the critical Cd content related to ecotoxcological impacts on soil organisms living in the soil (see De Vries et al., this volume).

To illustrate the impact of major soil types, critical total soil metal concentrations have been calculated on the basis of drinking water criteria for Cd for the following three soil types:

- Sandy soil (3% organic matter, 3% clay, and a pH H_2O of 5.5 in agriculture and a pH H_2O of 4.0 in nature)
- Clay soil (3% organic matter, 25% clay, and a pH H_2O of 6.5 in agriculture and a pH H_2O of 6.0 in nature)
- Peat soil (30% organic matter, 15% clay, and a pH H_2O of 6.0 in agriculture and a pH H_2O of 4.0 in nature)

Results thus obtained for those soil types illustrate that the lowest critical Cd and Pb contents are calculated for sandy soils in nonagricultural areas (Table 14). A discussion on the relevance of such limits is continued in Section IV.

III. Critical Soil Metal Concentrations Related to Impacts on Animal Health

Approach

Figure 5 shows the schematic representation of the link between acceptable daily intake (ADI) of animals and critical soil metal concentrations, distinguishing between birds, feeding on worms only, and mammals, feeding on birds and plants. Bioaccumulation of chemicals from soil to worm-eating birds and mammals takes place in at least two steps: first, the transfer from soil to food (plants and/or invertebrates; usually based on a BCF), which is followed by the transfer from food to higher organisms (small birds and mammals) using a BAF. The food chain of soil → plant (grass) → cattle has been described in our previous chapter for agricultural soils. This food chain is also relevant for grazing cows and sheep. In this case, the parameterization of the model is slightly different, but the overall result is comparable to that presented in Table 10 (De Vries et al. 2006). In this section, we focus on the food chain of soil → soil invertebrate → mammal/bird.

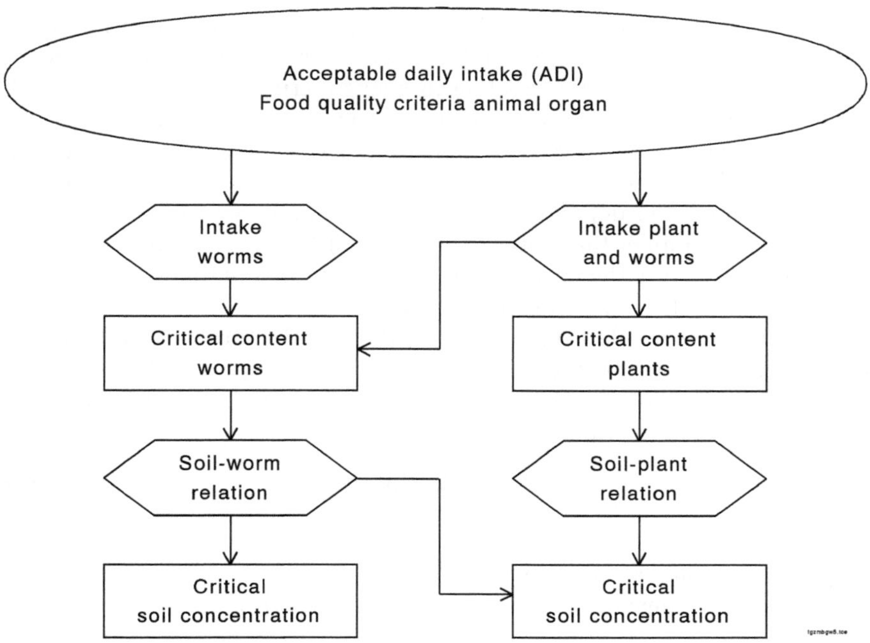

Fig. 5. Indicator and target organism and procedure that has been applied to derive critical limits for heavy metals in the soil from criteria in animal organs in view of toxic effects on animals.

Assuming that a mammal or bird feeds on soil invertebrates only, the simplest model to calculate a critical metal content in the soil, $[M]_{s(crit)}$, based on this food chain is the use of a bioconcentration factor, according to (Romijn et al. 1991a,b):

$$[M]_{s,tot(crit)} = [M]_{in(crit)} / BCF_{in} \tag{14}$$

in which:

$[M]_{in(crit)}$ = critical metal concentrations in terms of no observed effect concentrations (wet wt) of the food (invertebrate), corrected for the species of concern (mammal or bird: mg/kg)
BCF_{in} = bioconcentration factor, representing the ratio between the concentration in the invertebrate (the food of the species of concern) and the concentration in soil ($kg_{dry\ soil}/kg_{wet\ food}$)

The methodology has been used previously by Van de Plassche (1994) to derive critical soil metal contents for Cd, Cu, and methyl Hg, using the formula in the general sense of invertebrates, not only worms. Van de Plassche (1994) applied extra correction factors in Eq. 14 to extrapolate the results from toxicity studies in the laboratory to field conditions. These correction factors refers to differences in metabolic rate, caloric food content, food assimilation efficiency, pollutant assimilation efficiency, and species sensitivity to the pollutant in the laboratory and in the field situation. BCF values used by Van de Plassche (1994) have, however, not been corrected for soil characteristics, thus leading to a single critical limit value for Cd, Cu, and methyl Hg for all soils. A more sophisticated approach is the use of a BCF, which depends on soil characteristics, comparable to that of the soil–plant relationship as presented by Ma and van der Voet (1993) for Cd in earthworms. The dependence of critical metal contents in soil on soil characteristics implies that impacts of Cd on earthworms occur through the soil solution because the partitioning of Cd from the soil to the soil solution is influenced by the same soil characteristics.

As with the soil–plant relationships, the metal content in earthworms can, however, better be related to the metal content in soil in a nonlinear way, while accounting for the impact of soil properties. In this study we used such an approach (compare Eq. 2):

$$[M]_w = K_{sw} \cdot [M]_{s,tot}^{m} \tag{15}$$

where:

$[M]_w$ = metal concentration in worm (mg/kg)
K_{sw} = transfer constant from soil to worm ($[kg/mg]^m$)

in which the value of K_{sw} depends on the cation-exchange capacity (CEC) and the soil pH according to (compare Eq. 3, after Ma 1983):

$$\text{Log } K_{sw} = a_0 + a_1 \cdot \log(CEC) + a_2 \cdot pH \tag{16}$$

where:

CEC = cation-exchange capacity ($mmol_c$/100g)

By combining Eqs. 15 and 16, a critical soil limit can thus be calculated from an ADI using an inverse nonlinear soil–worm relationship according to:

$$[M]_{s,tot(crit)} = ([M]_{w,crit}) / K_{sw})^{1/m} \quad (17)$$

where:

$[M]_{w,crit}$ = critical limit for metal concentration in worm (mg/kg)

Application of the data of Ma (1983) to Eq. 15 and 16 for Cd and Pb, while deriving the CEC from the clay and organic matter content according to Helling et al. (1964), resulted in parameter values presented in Table 15 with a rather close match between data for internal levels of both metals in worms and model fit.

Below we further describe the approach to calculate critical metal concentrations in soil from critical metal contents in target organs and acceptable daily intakes, distinguishing between vermivores, feeding on worms only, and omnivores, feeding on both plants and worms.

For vermivores, the intake of earthworms is considered to be the dominant source of metals. Available information on the ADI of such a vermivore can be used to derive a critical metal content in the earthworm (the food) according to:

$$[M]_{w(crit)} = ADI / I_w \quad (18)$$

where:

I_w = Daily intake of earthworms (kg/d)

Equation 18 is based on the assumption that the vermivore eats earthworms only. Direct information on the acceptable daily metal intake is generally not available. However, this information can be derived from a critical metal content in the kidney of the vermivore and the critical time period in which this critical content is reached according to De Vries et al. (2007):

Table 15. Overview of Parameters in the Transfer Function for Metal Accumulation in Earthworms, Based on Data by Ma (1983).

	Parameters				
Metal	a_0	a_1 (CEC) $mmol_c$/100 g	a_2 (pH)	n mg/kg	R^2
Cd	2.69	−0.38	−0.14	0.51	0.72
Pb	1.92	−0.99	−0.22	1.16	0.61

$$[M]_{org(crit)} = [M]_{w(crit)} \cdot I_w \cdot f_{ass,org} \cdot T_{dy} \cdot T_{crit} / W_{org} \tag{19}$$

which by combination with Eq. 18 leads to

$$ADI = \frac{M_{org(crit)} \cdot W_{org}}{f_{ass,org} \cdot T_{dy} \cdot T_{crit}} \tag{20}$$

where

$[M]_{org\,(crit)}$ = critical limit for metal content in target organ (kidney) (mg/kg)
W_{org} = dry weight of the organ (kg)
$f_{ass,org}$ = assimilation fraction of the metal in food to the target organ (–)
T_{dy} = number of days during the year that the species is exposed to polluted food (d/yr)
T_{crit} = critical time period (reproductive phase of the species), in which the metal content in the target organ should stay below the critical limit (yr)

The kidney is used because this is the most sensitive organ for the intake of Cd, Pb, and Hg. The critical time period is set equal to the reproductive phase of the species.

When information on the ADI of an omnivore is available, this can be used to derive a critical metal content in the earthworm and the plant according to:

$$Ip \cdot [M]_{p(crit)} + Iw \cdot [M]_{w(crit)} = ADI \tag{21}$$

A combination of Eqs. 2, 16, and 21 leads to:

$$Ip \cdot K_{sp} \cdot [M]^{n}_{s,tot(crit)} + Iw \cdot K_{sw} \cdot [M]^{m}_{s,tot(crit)} = ADI \tag{22}$$

Equation 22 is based on the implicit assumption that the omnivore lives on one type of plant only. In principle, relationships have to be derived for all plant species that are a significant part of the diet of the omnivore. From Eq. 22, the value of $[M]_{s(crit)}$ can be solved iteratively on the basis of a given ADI and given values of K_{sp}, K_{sw}, Ip, and Iw. When a significant soil–plant relationship does not exist, a constant plant metal content (e.g., a median or 95th percentile value) should be used to calculate the soil content, according to:

$$[M]_{s,tot(crit)} = ((ADI - Ip \cdot [M]_p)/(Iw \cdot K_{sw}))^{1/m} \tag{23}$$

As with the vermivores, the value of ADI can be derived from a critical metal content in the kidney of the omnivore and the critical time period in which this critical content is reached, using Eq. 20.

Example of a Model Application

Below we illustrate the approach using data for soil–plant and soil–worm relationships and available target values for the kidney. The black-tailed

Table 16. Calculated Acceptable Daily Intake of Cd and Pb by the Black-Tailed Godwit and the Badger.

Animal	$[M]_{org(crit)}$ (mg/kg)		M_{org} (kg)	$f_{ass,org}$ (–)		T_{dy} (d/yr)	T_{crit} (yr)	ADI (mg/d)	
	Cd	Pb		Cd	Pb			Cd	Pb
Godwit[a]	200[c]	90[d]	3.85×10^{-3}	5×10^{-3}	1.5×10^{-4}	122	5	0.253	0.114
Badger[b]	200[c]	90[d]	65×10^{-3}	5×10^{-3}	1.5×10^{-4}	365	4	1.781	0.801

[a]Apart from the critical Cd and Pb contents in the kidney, $[M]_{org(crit)}$, all data are based on Bosveld et al. (2000).

[b]Apart from the critical Cd and Pb contents in the kidney, all data are based on Klok et al. (1998).

[c]The critical limit of Cd in the kidney of vertebrates varies is based on a LOEC of 100–350 mg/kg (Nicholson et al. 1983; Cooke and Johnson 1996; Pascoe et al. 1996). In this chapter, we used an intermediate value of 200 mg/kg.

[d]This critical limit is based on Ma (1996).

godwit, a small bird, was taken as a representative of the vermivores and the badger was chosen as a representative of the omnivores. For the badger, the intake of earthworms (*Lumbricus terrestris*) forms the largest part of their diet, for which well-grazed pastures are preferred. Badgers, however, also eat grass, fruits, and nuts, cereals such as wheat or oats, bulbs and tubers, etc. In short, badgers are opportunists and will consume whatever is available, but earthworms are the preferred food item. To illustrate the approach, we assumed that the badger lives on worms and grass only. The calculation of critical soil metal concentrations has been limited to Cd and Pb, as information for Hg needed to calculate ADI values and critical metal contents in worms was not available.

Estimates of the ADI, as well as the parameters required to perform the calculation, are given in Table 16. From the ADI values, the critical metal content in soil was calculated assuming an intake of worms (wet wt) of 0.1 kg/d by the godwit and 0.5 kg/d by the badger and a dry matter percentage for worms of 16%. The intake of plant material by the badger was also set at 0.5 kg/d. Table 16 also includes results for the badger, based on the assumption that they feed on worms only (worst case situation; compare Eqs. 17 and 23).

Results of the critical metal concentrations for cadmium and lead in soil based on ADIs of those metals by the godwit and badger, determined by the target values for those metals in the kidney, are given in Table 17. A distinction has been made between agricultural and nonagricultural soil based on the expected difference in soil pH. With respect to clay and organic matter content, use was made of the values presented earlier. The pH values used are the following:

Table 17. Overview of Critical Total Cd and Pb Concentrations in the Soil Based on Acceptable Daily Intakes of Those Metals by the Black-Tailed Godwit and Badger.

		Critical Cd content (mg/kg)		Critical Pb content (mg/kg)	
Soil use	Soil type	Black-tailed godwit	Badger	Black-tailed godwit	Badger
Agriculture	Sand	0.14	0,26 (0,28)	123	157 (165)
Agriculture	Clay	0.66	1,2 (1,3)	534	668 (718)
Agriculture	Peat	1.0	1,9 (2,0)	1024	1297 (1378)
Nature	Sand	0.067	0,12 (0,13)	69	88 (92)
Nature	Clay	0.47	0,82 (0,92)	412	514 (554)
Nature	Peat	0.33	0,60 (0,65)	426	539 (573)

Data for the badger are based on a daily intake of 0.5 kg worms and 0.5 kg plant material.

Values in brackets are based on a daily intake of worms only.

- Sandy soil: 5.5 for agriculture and 4.5 for nature
- Clay soil: 6.5 for agriculture and 6.0 for nature
- Peat soil: 6.0 for agriculture and 4.5 for nature

Results show that calculated critical soil Cd concentrations are very low, especially on sandy soils (see Table 17). Calculated values are (much) lower than those based on ecotoxicological criteria (see De Vries et al., this volume). Often, the values are also below present Cd concentrations in soils, which implies a present risk for these worm-eating mammals and birds. In contrast to Cd, critical Pb concentrations are high (Table 17) and far above, up to ten times, the critical concentrations related to ecotoxicological impacts (see De Vries et al., this volume) and the generally observed present Pb concentrations.

IV. Discussion

A. Uncertainties in Deriving Critical Soil Metal Concentrations from Food Quality Criteria for Crops

Uncertainties in Food Quality Criteria

The derivation of critical soil metal concentrations based on critical content in crops is based on the idea that products from agricultural soils have to meet standards for food. The choice of the standard has a profound impact on the level of the critical metal content in soils., in case of a significant soil–plant relationship, as derived for Cd; this holds specifically for wheat, which is a dominant source of Cd exposure to humans. The use of the recommended food quality criterion of 0.20 mg/kg causes a large difference in critical soil Cd concentration compared to the formerly used value of 0.10 mg/kg. Many studies mention considerable gaps in knowledge, particularly in assessing the risk to human health from exposure to dietary Cd. On one hand, new scientific results (Ikeda et al. 2003; Simmons et al. 2003; Chaney et al. 2004; Reeves and Chaney 2004; Reeves et al. 2005) induced discussions whether even the currently used official European critical Cd limits of 0.2 mg/kg (fresh weight) for cereals might possibly be too low. Based on a meta-analysis of human surveys in Japan, Ikeda et al. (2003) presented evidence on a threshold in urinary Cd before renal tubular dysfunction occurred that was clearly higher than thresholds reported in European studies. The latter were, however, not related to evidence of Cd disease but of predisease conditions. On the other hand, there are indications that the 0.10 mg/kg criterion is already too high. In the derivation of a Cd ADI for adult nonsmokers of 37–47 μg (EC 2003), an absorption rate of dietary Cd by the human body of only 3% was assumed. The Cd absorption rates by the human body include some uncertainties and, if a higher absorption rate were to be applied, e.g., the frequently used value of 5% (Kalberlah 1999), the derived ADI would be lower, thus lowering the acceptable Cd level in grain. Furthermore, the EC CSTEE (2004) stated that sensitive parts of the population are not sufficiently considered in the RAR-Cd and recommended the use of more conservative approaches for risk assessment in general. This precautionary approach is in line with statements from the WGE (2004), the EC DG Industry (1997), JECFA (2000), and SCOPE (2003) that there is no safe level for Cd in food.

Uncertainties in Soil–Plant Relationships for Metals

Several aspects play an important role when deriving soil to plant relationships to be used for the derivation of the soil critical metals concentrations:

1. *Impact of plant species* (variation between cultivars as well as crops). The uptake of metals is known to vary between crops. In general, crops such as lettuce and wheat tend to accumulate more metals than crops such as beans and tomatoes. Thresholds for agriculture should therefore be based on the more sensitive crops. Once the criteria for sensitive crops are met, the cultivation of other crops is secured. Variations in uptake of metals between cultivars, however, also play a considerable role. In the data used for the derivation of the soil to plant relationships for the Netherlands, several cultivars were planted, this being one reason why the quality of soil to plant relationships is often rather poor.

2. *Validity of data for specific conditions*. The availability of metals in soils, and hence the uptake of metals by crops, depends on soil conditions

and soil properties such as clay content, organic matter, and pH. Differences in type of clay minerals and organic matter quality in soils between different climatic regions may cause different uptake patterns. Also, the uptake of water is rather different in soils in semiarid regions compared to those in moderate climates. The data used here are likely to be representative for conditions that prevail in northwestern parts of Europe, but care should be taken using plant to soil relationships based on these data in other climatic zones. When looking for data to derive critical limits in soil, care should be taken that these data should cover the range of interest in both soil properties as well as the metal content in soils and crops. The highest plant metal content should at least be equal to the food quality standard chosen to avoid extrapolation of the regression equations.

3. *Use of data from experiments performed using hydroponic solutions and in greenhouses.* Differences in conditions between experiments (in greenhouses and lysimeters) and those in the field will cause different uptake patterns of metals from soils. In general, growing conditions in most plant uptake experiments are kept constant. Plants are watered frequently, and the nutrient status is often maintained by adding ample supplies of fertilizer, which will affect the uptake of metals from these soils. To obtain critical limits valid at field conditions, data from field experiments should therefore be used.

4. *Availability of data for specific combinations of metals and crops.* For Cd, Pb, and Hg the number of data for the crops of interest is rather large due to the extensive monitoring of soil and crops by Wiersma et al. (1986). However for other metals such as Cu and Zn, no such national inventories are yet available.

5. *Use of "historic" data to derive soil to plant relationships.* Most data used in this study are based on monitoring efforts during the early 1980s. After that time, atmospheric deposition of lead, especially, has decreased considerably because of the shift to lead-free fuel. Because Pb uptake is related to atmospheric deposition, plant data from the pre-lead-free fuel era might not be applicable to current conditions. The fact that little or no relationship between soil and plants for lead could be obtained may be partially related to the fact that the plant metal level was controlled (partly) by atmospheric deposition. Differences in atmospheric deposition usually are less pronounced than differences in the soil Pb levels and soil properties across the Netherlands.

Despite some obvious points of concern raised above, the model concept presented here is already a step ahead compared to the common concept used (BSF). It has been shown that uptake of metals by crops can be described by nonlinear equations more accurately compared to linear bioaccumulation factors. Differences in critical levels between soil types are large and need to be considered when trying to derive relevant protection levels for different soil types. At present, more complex models exist to predict plant uptake and heavy metal availability in soil. However, these models require a rather extensive parameterization that is, usually, hard to obtain for application in different soil types. An advantage of the approach described here is that it can be easily applied to a regional or even national scale because the input required is usually available.

B. Uncertainties in Deriving Critical Soil Metal Concentrations from Food Quality Criteria for Animal/Products

Uncertainties in the derivation of critical soil concentrations from critical metal contents in animal organs/products from cows and sheep are determined by uncertainties in the acceptable daily intake (ADI), the daily intake of plant and soil, and in the transfer rates of metals from soil to grass to the consuming animal. Uncertainties in the intake of plant and soil are comparatively limited. Regarding soil–plant relationships, the uncertainties have already been mentioned. Most uncertain, however, is the hypothesized linear relationship between metal content in animal organs/products and metal content in fodder. The use of a constant bioaccumulation factor from plant to animal organ, BAF_{pa}, is a crucial assumption in deriving ADI values from critical metal contents in animal organs/products (see Eq. 10). Schütze et al. (2003a) made a literature review on the relationships between Cd in environment/fodder and Cd in animals (organs and muscle of wildlife and cattle). In several studies (Hapke et al. 1977; Crössmann 1981; Schinner 1981; Hecht 1982; Holm 1983), it could be shown that there is a relationship. The overall conclusion was, however, that a mathematical quantification of the carry-over could not be done.

The carry-over rates to a certain organ depend not only on the Cd intake but also on animal species, animal age, and the composition of fodder. In particular, with respect to Cd, there is a strong interlink to Zn uptake. We also must be aware that the differences in metal contents in fodder are usually low compared with the gradients in studies and surveys reported in the literature. The correlation between metal in fodder and in animal product is probably much weaker in cases of unpolluted soils. Another crucial assumption is that the availability of metals present in plant products and soil to animals is the same. This assumption also needs verification. In summary, when ADI values are not available, the derivation of a soil critical concentration is highly uncertain.

C. Uncertainties in Deriving Critical Soil Metal Concentrations from Drinking Water Quality Criteria

Uncertainties in Transfer Functions from Soil to Soil Solution

The points raised in relation to the soil to plant relationships are to a large extent also valid for the soil to water relationships. Data (especially the range therein) used to construct the relationships should match the range in environmental conditions. Again, differences in geology and specific sources of contamination (nature of the contamination) have a profound impact on the distribution of metals between soil and soil solution. As such, the approach outlined here is merely a conceptual approach (valid for the Netherlands, of course), and care should be taken when using the relationships presented here under rather different circumstances.

Using a Critical Metal Concentration in Soil Solution Based on Drinking Water Quality Criteria

An important methodological aspect is the fact that critical soil concentrations calculated from groundwater standards (based on drinking water quality criteria) assume that the metal concentration in soil solution can be used as an estimate of the concentration in groundwater. This is only the case in a steady-state situation, assuming a constant dissolved metal concentration with depth. Because steady state is generally not attained, the approach presented here is a worst case. At a given critical soil concentration, the water draining from the topsoil, for which the calculations are made, will generally contain more metals than upper groundwater because of ongoing retention in the lower parts of the soil profile. Only a dynamic modeling approach can account for these changes in the soil profile. In this way, one can calculate the time period before the dissolved metal concentration reaches a critical value in groundwater. Such a calculation should be made using the critical metal input (critical load) to the soil, accounting for differences between the critical and original soil metal concentration at each soil depth until groundwater, is reached. When this time period is longer than a target period (e.g., 100 years), one may not want to use such limits for regulation purposes.

D. Uncertainties in Deriving Critical Soil Metal Concentrations from Critical Metal Contents in Organs of Worm-Eating Birds

The assessment of critical soil metal concentrations from critical metal contents in organs of worm-eating birds is influenced by the uncertainty in all the factors affecting the calculation, including the ADI of the worm-eating animal, the intake of worms, and the soil to worm relationships. Data on the intake of worms are comparatively reliable. As with the soil–plant relationships, the data used to construct soil–worm relationships are specific for Dutch circumstances with respect to clay mineralogy, organic matter quality, etc., but in these circumstances, they seem quite reliable. As with the calculation for cows and sheep, the largest uncertainty is related to the acceptable daily intake. The uncertainty of the ADI value is, among others, determined by the uncertainty in critical internal level for metal in the target organ of the worm-eating animal and the assimilation fraction of the metal in food to the target organ (see Eq. 20). Both uncertainties are relatively large. In addition, it is important to mention that the back-calculated soil concentrations assumes a homogeneous feeding habit area of the animal. Because animals move around, the quality of the food they consume differs from one location to the other, depending on local conditions. The size of the feeding area also affects to what extent this difference affects the exposure of animals. The earthworms will remain bound to a certain area, but birds and mammals, such as the badger, that are feeding on the worms dwell in a much larger area, and the variability in the degree of contaminants (spatial variation in the amount of metals in soil, for example), will affect the levels of metals present in the food obtained from one site compared to another. This aspect is spatially averaged in back-calculating the concentrations.

V. Conclusions

Major conclusions that can be derived from the presented analyses in this chapter follow.

1. The impact of soil properties on critical soil metal concentrations in view of human health and animal health impacts is mainly relevant for Cd, because of the occurrence of rather significant soil–plant, soil–solution, and soil–worm relationships. For Hg, the effects are unclear and presently cannot be included. There are no soil–soil solution, soil–plant, or soil–worm relationships. Only in grazing animals is a back-calculation possible because of ingestion of contaminated soil as the dominant pathway. However, this calculation is highly uncertain. For Pb, impacts of soil properties can be relevant in case of worm-eating animals and in view of impacts on groundwater quality from the occurrence of rather significant soil–solution and soil–worm relationships. Critical soil concentrations for Pb related to impacts on animal health and human health are, however, generally much higher than those related to ecotoxicological impacts (the same is true for Hg).

2. The largest uncertainties in deriving critical soil concentrations in view of human health and animal health are related to soil concentrations that are derived from food quality criteria or internal critical limits of animal organs; this is mainly because the metal transfer rates from soil to food (either plants such as grass or small animals such as worms) to animals vary considerably, and model approaches are still in development. Both the availability of the metal in the food or soil consumed by animals and the actual internal uptake by animals are concerns. Furthermore, there is a large uncertainty in the internal critical level in organs for both aboveground and belowground organisms (see Bruus Pedersen et al. 2000).

3. Critical Cd concentrations in view of health effects on animals and humans are sometimes lower than those related to ecotoxicological impacts on soil organisms/processes and plants. This point is illustrated in Table 18,

Table 18. Calculated Critical Total Cd Contents in Soil in View of Impacts on Soil Organisms/Soil Processes, Food Quality of Wheat, and Health of Worm-Eating Birds and Mammals.

		Critical Cd content (mg/kg)		
Land use	Impact on	Sand	Clay	Peat
Nature[a]	Soil organisms (pH 4.5)	0.89	2.8	13
Arable land	Wheat	0.46	0.72	1.9
Nature[a]	Drinking water	0.24	1.2	1.4
Nature[a]	Impacts on worm eating mammals (badger)	0.13	0.92	0.65
Nature[a]	Impacts on worm-eating birds (godwit)	0.067	0.47	0.33

[a]Apart from the critical Cd and Pb contents in the kidney, $[M]_{org(crit)}$, all data are based on Bosveld et al. (2000).

showing calculated critical total Cd concentrations in soil related to food quality criteria for wheat, drinking water quality, and acceptable daily intakes of worm-eating birds and mammals. Specifically, for acid sandy soils the calculated critical concentrations are (much) lower than critical metal concentrations related to ecotoxicological impacts (see also De Vries et al., this volume). Despite the uncertainties involved, this implies that present Cd concentrations in the rural area may affect both agricultural and non-agricultural systems.

Summary

Assessment of the risk of elevated soil metal concentrations requires appropriate critical limits for metal concentrations in soil in view of ecological and human toxicological risks. This chapter presents an overview of methodologies to derive critical total metal concentrations in soils for Cd, Pb, and Hg as relevant to health effects on animals and humans, taking into account the effect of soil properties. The approach is based on the use of nonlinear relationships for metals in soil, soil solution, plants, and soil invertebrates, including soil properties that affect metal availability in soil. Results indicate that the impact of soil properties on critical soil metal concentrations is mainly relevant for Cd because of significant soil–plant, soil–solution, and soil–worm relationships. Critical Cd levels in soil thus derived are sometimes lower than those related to ecotoxicological impacts on soil organisms/processes and plants, which is especially true for critical soil Cd concentrations in view of food quality criteria for wheat, drinking water quality, and acceptable daily intakes of worm-eating birds and mammals. There are, however, large uncertainties involved in the derivation from assumptions made in the calculation and uncertainties in acceptable daily intakes and in relationships for Cd in soil, soil solution, plants, and soil invertebrates. Despite these uncertainties, the analyses indicate that present Cd concentrations in parts of the rural areas are in excess of the critical levels at which effects in both agricultural and nonagricultural systems can occur.

Acknowledgments

This work was funded by the Ministry of Housing, Spatial Planning and the Environment and the Ministry of Agriculture, Nature and Food Quality of the Netherlands, and the UK Department of the Environment, Food and Rural Affairs under grant EPG /1/3/188.

References

Anonymous (2000) Health risks of heavy metals from long range transboundary air pollution. EB.AIR/WG.1/2000/12. Geneve, UN-ECE.

Adams ML, Zhao FJ, McGrath SP, Nicholson FA, Chambers BJ (2004) Predicting cadmium concentrations in wheat and barley grain using soil properties. J Environ Qual 33:532–541.

Adriano DC (2001) Trace Elements in Terrestrial Environments: Biogeochemistry, Bioavailability, and Risks of Metals. Springer, New York.

Beresford N, Mayes RW, Crout NMJ, MacEachern PJ, Dodd BA, Barnett C, Stuart Lamb C (1999) Transfer of cadmium and mercury to sheep tissues. Environ Sci Technol 33:2395–2402.

BgVV (Bundesinstitut für gesundheitlichen Verbraucherschutz und Veterinärmedizin) (1997) Lebensmittelmonitoring 1997. Zentrale Erfassungs- und Bewertungsstelle für Umweltchemikalien (ZEBS), Berlin, Germany.

Bosveld ATC, Klok C, Bodt JM, Rutgers M (2000) Ecologische risico's van bodemverontreinigingen in toemaakdek in de gemeente De Ronde Venen. Alterra rapport 151. Alterra, Research Instituut voor de Groene Ruimte, Wageningen, Nederland.

Brus DJ, de Gruijter JJ, Walvoort DJJ, de Vries F, Bronswijk JJB, Römkens PFAM, de Vries W (2002) Mapping the probability of exceeding critical thresholds for cadmium concentrations in soils in the Netherlands. J Environ Qual 31:1875–1884.

Bruus Pedersen M, van Gestel CAM, Elmegaard N (2000) Effects of copper on reproduction of two collembolan species exposed through soil, food and water. Environ Toxicol Chem 19:2579–2588.

Chaney RL, Reeves PG, Ryan JA, Simmons RW, Welch RM, Angle JS (2004) An improved understanding of soil Cd risk to humans and low cost methods to phytoextract Cd from contaminated soils to prevent soil Cd risks. BioMetals 17:549–553.

Chang AC, Granato TC, Page AL (1992) A methodology for establishing phytotoxicity criteria for chromium, copper, nickel, and zinc in agricultural land application of municipal sewage sludges. J Environ Qual 21:521–536.

Cooke JA, Johnson MS (1996) Cadmium in small mammals. In: Beyer WN, Heinz GH, Redman AW (eds) Environmental Contaminants in Wildlife: Interpreting Tissue concentrations. Lewis, New York, USA, pp 377–388.

Crommentuijn T, Polder MD, van de Plassche EJ (1997) Maximum permissible concentrations and negligible concentrations for metals, taking background concentrations into account. RIVM Report 601501 001. National Institute of Public Health and the Environment, Bilthoven, The Netherlands.

Crössmann G (1981) Ein Beitrag zur Aufnahme und Rückstandsbildung von Cadmium bei Mastschweinen. Arch Lebensmittelhyg 1:21–24.

De Temmerman L, de Witte T (2003a) Biologisch onderzoek van de verontreiniging van het milieu door zware metalen te Hoboken groeiseizoen 2001. Internal report Centrum voor onderzoek in diergeneeskunde en agrochemie. C.O.D.A, Tervuren, België.

De Temmerman L, de Witte T (2003b) Biologisch onderzoek van de verontreiniging van het milieu door kwik en chloriden te Tessenderlo en van kwik te Berendrecht. Internal report Centrum voor onderzoek in diergeneeskunde en agrochemie. C.O.D.A, Tervuren, België.

De Vries W, Schütze G, Lofts S, Tipping E, Meili M, Römkens PFAM, Groenenberg JE (2005) Calculation of critical loads for cadmium, lead and mercury. Background document to a mapping manual on critical Loads of cadmium, lead and mercury. Alterra Report 1104. Alterra, Wageningen, The Netherlands.

De Vries W, Römkens PFAM, Rietra RPJJ, Bonten L, Ma WC, Faber J, Harmsen J, Bloem J (2007) De invloed van bodemeigenschappen op kritische concentraties voor zware metalen en organische microverontreinigingen in de bodem. Alterra rapport 817. Alterra, Wageningen, The Netherlands.

Dijkshoorn W, van Broekhoven W, Lampe JEM (1979) Phytotoxicity of zinc, nickel, cadmium, lead, copper and chromium in three pasture plant species supplied with graduated amounts from the soil. Neth J Agric Sci 27:241–253.

EC (2003) Risk assessment report on cadmium metal and cadmium oxide (the targeted report on the use in batteries, included): final draft version (July 2003). European Commission. Available on the ECB web-site: http://ecb.jrc.it.

EC CSTEE (European Commission Scientific Committee on Toxicity, Ecotoxicity and the Environment) (2004) Opinion on the result of the risk assessment of cadmium metal human health, cadmium oxide human health. Adopted by the CSTEE on 08 Jan 2004.

EC DG Industry (1997) European Commission, Food Science and Techniques: Opinion of the Scientific Committee for Food on Cadmium expressed on 2 June 1995. Reports of the Scientific Committee for Food (36th series). European Commission Directorate-General Industry, Brussels, Belgium.

EU (2001) Verordening nr. 466/2001 van 8 maart 2001, tot vaststelling van maximumgehalten aan bepaalde verontreinigingen in levensmiddelen. Publicatieblad van de Europese Gemeenschappen L 77.

Friberg L, Kjellström T, Nordberg. G, Piscator M (1979) Cadmium. Elsevier/North-Holland, Amsterdam.

Hapke H-J, Abel J, Kühl U, Glaser U (1977) Die Ansammlung von Cadmium in verzehrbaren Geweben in Abhängigkeit von der Cadmium-Menge im Futter. Arch Lebensmittelhyg 28:174–177.

Hecht H (1982) Toxische Schwermetalle in Fleisch und Innereien verschiedener Tierarten. Mitteilungsbl Bundesanst Fleischforsch 76:5053.

Helling CS, Chesters G, Corey RB (1964) Contribution of organic matter and clay to soil cation exchange capacity as affected by the pH of the saturating solution. Soil Sci Soc Am J 28:517–520.

Holm J (1983) Aufbau eines ursachenorientierten Monitoring-Systems für Schadstoffbelastungen beim Wild. Fleischwirtschaft 63:1764–1766.

Huinink J (2000) Functiegerichte bodemkwaliteitssystematiek 2. Functiegerichte bodemkwaliteits waarden. IKC-Landbouw, Ede, Nederland.

Ikeda M, Ezaki T, Tsukahara T, Moriguchi J, Furuki K, Fukui Y, Ukai H, Okamoto S, Sakurai H (2003) Threshold levels of urinary cadmium in relation to increases in urinary β2-microglobulin among general Japanese populations. Toxicol Lett 137:135–141.

Jakubowski M (2003) Health risk from environmental exposure to heavy metals. Report prepared for the WHO-ECEH-UNECE Task Force on Health Aspects of Long Range Transboundary Air Pollution.

JECFA (Joint FAO/WHO Expert Committee on Food Additives) (2000) Fifty-fifth meeting. Summary and conclusion. Geneva, 6–15 June 2000.

Jongbloed RH, Pijnenburg J, Mensink BJWG, Traas TP, Luttik R (1994) A model for environmental risk assessment and standard setting based on biomagnification. Top predators in terrestrial ecosystems. RIVM Report 719101 012. National Institute of Public Health and the Environment, Bilthoven, The Netherlands.

Kabata-Pendias A, Pendias H (1992) Trace Elements in Soils and Plants. CRC press, Boca Raton, FL.

Kalberlah F (1999) Cadmium und seine Verbindungen. In: Basisdaten Toxikologie für umweltrelevante Stoffe zur Gefahrenbeurteilung bei Altlasten mit Ableitung von toxikologisch begründeten Werten resorbierten Körperdosen (TRD-Werten) Bericht zum F+E Vorhaben 103 40 113. Umweltforschungsplan des Ministeriums für Umwelt, Naturschutz und Reaktorsicherheit. Umweltbundesamt, Berlin, Germany.

Kawada T, Suzuki S (1998) A review on the cadmium content of rice, daily cadmium intake, and accumulation in the kidneys. J Occup Health 40:264–269.

Klok C, de Roos AM, Broekhuizen S, van Apeldoorn RC (1998) Effecten van zware metalen op de Das. Interactie tussen versnippering en vergiftiging. Landschap 15:77–86.

Kübler W, Anders HJ, Heeschen W (eds) (1995) Bayerische Verzehrsstudie 1995. In: Ergebnisse der nationalen Verkehrsstudie 1985–88 über die Lebensmittel- und Nährstoffaufnahme in der BRD: Band XI. Wiss. Fachverlag Fleck. VERA Schriftenreihe, Niederkleen, Germany.

Lijzen JPA, Baars AJ, Otte PF, Rikken MGJ, Swartjes FA, Verbruggen EMJ, van Wezel AP (2001) Technical evaluation of the intervention values for soil/sediment and groundwater. Human and ecotoxicological risk assessment and derivation of risk limits for soil, aquatic sediment and groundwater. RIVM report 711701 023. Rijksinstituut voor Volksgezondheid en Milieu, Bilthoven, Nederland.

Ma W, van der Voet H (1993) A risk-assessment model for toxic exposure of small mamalian carnivores to cadmium in contaminated natural environments. Sci Total Environ Suppl 1993:1701–1714.

Ma WC (1983) Regenwormen als bio-indicators van bodemverontreiniging. Min. van VROM. Staatsuitgeverij, Den Haag, Nederland.

Ma WC (1996) Lead in mammals. In: Beyer WN, Heinz GH, Redmon-Norwood AW (eds) Environmental Contaminants in Wildlife: Interpreting Tissue Concentrations. Lewis, New York, pp 281–296.

Ma WC, Bosveld ATC, van den Brink DB (2001) Schotse hooglanders in de broekpolder? Analyse van de veterinair-toxicologische risico's van de verontreinigde bodem voor grote grazers. Alterra rapport 260. Alterra, Research Instituut voor de Groene Ruimte, Wageningen, Nederland.

MacNicol RD, Beckett PHT (1985) Critical tissue concentrations of potentially toxic elements. Plant Soil 85:107–129.

McKone TE, Ryan PB (1989) Human exposures to chemicals through food chains: an uncertainty analysis. Environ Sci Technol 23:1154–1163.

Meili M (1997) Mercury in lakes and rivers. In: Sigel A, Sigel H (eds) Mercury and Its Effects on Environment and Biology. Dekker, New York, pp 21–51.

Meili M, Bishop K, Bringmark L, Johansson K, Munthe J, Sverdrup H, de Vries W (2003) Critical levels of atmospheric pollution: criteria and concepts for operational modelling of mercury in forest and lake ecosystems. Sci Total Environ 304:83–106.

Mortvedt JJ, Cox FR, Shuman LM, Welch RM (1991) Micronutrients in agriculture. Soil Science Society of America (SSSA), Madison, WI.

Nicholson JK, Kendall MD, Osborn D (1983) Cadmium and mercury nephrotoxicity. Nature (Lon) 304:633–635.

Pascoe GA, Blanchet RJ, Linder G (1996) Food chain analysis of exposure and risks to wildlife at a metals-contaminated wetland. Arch Environ Contam Toxicol 30:306–318.

Puls R (1988) Mineral Levels in Animal Health: Diagnostic Data. Sherpa International, Clearbook, BC, Canada.

Reeves PG, Chaney RL (2004) Marginal nutritional status of zinc, iron, and calcium increases cadmium retention in the duodenum and other organs of rats fed rice-based diets. Environ Res 96:311–322.

Reeves PG, Chaney RL, Simmons RW, Cherian MG (2005) Metallothionein induction is not involved in cadmium accumulation in the duodenum of mice and rats fed diets containing high-cadmium rice or sunflower kernels and a marginal supply of zinc, iron, and calcium. J Nutr 135:99–108.

Rikken MGJ, Lijzen JPA, Cornelese AA (2001) Evaluation of model concepts on human exposure. Proposals for updating the most relevant exposure routes of CSOIL. RIVM report 711701 022. National Institute of Public Health and the Environment, Bilthoven, The Netherlands.

Romijn CAF, Luttik R, van de Meent D, Slooff W, Canton JH (1991a) Presentation and analysis of a general algorithm for risk-assessment on secondary poisoning. RIVM Report 679102 002. National Institute of Public Health and the Environment, Bilthoven, The Netherlands.

Romijn CAF, Luttik R, van de Meent D, Slooff W, Canton JH (1991b) Presentation of a general algorithm for effect-assessment on secondary poisoning. RIVM Report 679102 007. National Institute of Public Health and the Environment, Bilthoven, The Netherlands.

Römkens PFAM, Groenenberg JE, Bonten LTC, de Vries W, Bril J (2004) Derivation of partition relationships to calculate Cd, Cu, Ni, Pb and Zn solubility and activity in soil solutions. Alterra Rapport 305. Alterra, Wageningen, The Netherlands.

Satarug S, Haswell-Elkins MR, Moore MR (2000) Safe levels of cadmium intake to prevent renal toxicity in human subjects. Br J Nutr 84:791–802.
Sauerbeck D (1983) Welche Schwermetallgehalte in Pflanzen dürfen nicht überschritten werden, um Wachstumsbeeinträchtigungen zu vermeiden? Landwirtsch Forsch Sonderh 39:108–129.
Schinner L (1981) Untersuchungen über endogene und exogene Einflüsse auf den Blei-(Pb-) und Cadmium- (Cd-) Gehalt in Muskeln und Organen von Rehwild und Wildkaninchen. Mitt Bund Fleischforsch 75:3056.
Schütze G, Becker R, Dämmgen U, Nagel H-D, Schlutow A, Weigel H-J (2003a) Risikoabschätzung der Cadmium-Belastung für Mensch und Umwelt infolge der Anwendung von cadmiumhaltigen Düngemitteln. Landbauforsch Völkenrode 2/3 2003:63–170.
Schütze G, Lorenz U, Spranger U (2003b) Expert Meeting on Critical Limits for Heavy Metals and Methods for Their Application. Proceedings, Berlin, 2–4 December 2003, UBA Texte 47/2003. Umweltbundesamt, Berlin, Germany.
SCOPE (Scientific Committee on Problems of the Environment) (2003) Risk Assessment and Management of Environmental Cadmium. Report from the SCOPE Workshop, University of Ghent, Belgium, 3–6 September, 2003.
Sheppard SC (1992) Summary of phytotoxic levels of soil arsenic. Water Air Soil Pollut 64:539–550.
Simmons RW, Pongsakul P, Chaney RL, Saiyasitpanich D, Klinphoklap S, Nobuntou W (2003) The relative exclusion of zinc and iron from rice grain in relation to rice grain cadmium as compared to soybean: implications for human health. Plant Soil 257:163–170.
Smilde KW (1976) Toxische gehalten aan zware metalen (Zn, Cu, Cr, Ni, Pb, en Cd) in grond en gewas: een literatuuroverzicht. Nota Instituut voor Bodemvruchtbaarheid no 25, I.B., Haren, Nederland.
Statistisches Bundesamt (1999) Verbrauch an Nahrungsmitteln in Deutschland 1991 bis 1998. In: Statistisches Jahrbuch über Ernährung, Landwirtschaft und Forsten der Bundesrepublik Deutschland. Metzler-Poeschel, Stuttgart, Germany.
Sverdrup H (2002) Setting critical limits for mercury, cadmium and lead to be used in calculation of critical loads for different receptors. Internal report. Chemical Engineering, Lund University, Lund, Sweden.
Van de Plassche EJ (1994) Towards integrated environmental quality objectives for several compounds with a potential for secondary poisoning. Report 679101 012. National Institute of Public Health and the Environment, Bilthoven, The Netherlands.
Van den Berg R, Roels JM (1991) Risk assessment to man and the environment in case of exposure to soil contamination. Integration of different aspects. RIVM report 725201 013. Rijksinstituut voor Volksgezondheid en Milieuhygiëne, Bilthoven, The Netherlands.
Van Driel W, van Luit B, Schuurmans W, de Vries W, Stienen MJJ (1988) Zware metalen in oevergronden en daarop verbouwde gewassen in het stroomgebied van Maas, Geul en Roer in de provincie Limburg. 3. Bodem-gewasrelaties. Rapport van de projectgroep zware metalen in oevergronden van Maas en zijrivieren.
Van Hooft WF (1995) Risico's voor de volksgezondheid als gevolg van blootstelling van runderen aan sporenelementen bij beweiding. Rapport 693810 001. Rijksinstituut voor Volksgezondheid en Milieuhygiëne, Bilthoven, Nederland.

VROM (1998) Soil Protection Act. Ministry of Housing, Spatial Planning and the Environment, The Hague, The Netherlands.

WGE (2004) Review and assessment of air pollution effects and their recorded trends. Report of the Working Group on Effects, Convention on Long-Range Transboundary Air Pollution. National Environment Research Council, UK.

WHO (2004) Guidelines for Drinking-Water Quality. Volume 1: Recommendations. WHO, Geneva.

Wiersma D, van Goor BJ, van der Veen NG (1986) Cadmium, lead, mercury and arsenic concentrations in crops and corresponding soils in the Netherlands. J Agric Food Chem 34:1067–1074.

Manuscript received July 21, accepted August 2, 2006.

Rev Environ Contam Toxicol 191:131–162

Fluoroquinolone Antibiotics in the Environment

Premasis Sukul and Michael Spiteller

Contents

I. Introduction

The occurrence of human and veterinary pharmaceuticals in soil and water (Thiele-Bruhn 2003; Snyder et al. 2003; Hamscher et al. 2004; Kay et al. 2004) has led to increased research activities among environmental scientists to find out their possible environmental threats. As antibiotics are used for human and animal medical care, there is a possibility for these drugs to reach the environment via direct or indirect contamination (Thiele-Bruhn 2003; Boxall et al. 2004). As they are produced and applied with the aim of being

Communicated by David M. Whitacre.

P. Sukul (✉), M. Spiteller
Institute of Environmental Research (INFU), Otto-Hahn-Strasse 6, University of Dortmund, 44221 Dortmund, Germany

biologically highly effective, their occurrence is of ecotoxicological interest. In Berlin, Germany, the groundwater wells located near contaminated surface waters showed a concentration of pharmaceuticals at the μg/L level (Heberer 2002). Thus, at such a level of contamination, the use of groundwater for drinking purposes may pose a potential risk. Studies on the behavior of animal drugs and their metabolites after excretion, along with their transport from agricultural sources into surface water and groundwater by overland-flow runoff and leaching, are of utmost importance at present.

Most environmental studies on pharmaceuticals are directed toward antibiotics due to emerging problem of antibiotic-resistant bacteria (Daughton and Ternes 1999). Being biocidal, the antibiotics may affect both bacteria and soil- and water-dwelling organisms when they reach the environment. The quinolones have been recognized as a potent group of antibiotics for the last four decades and are used to treat a variety of bacterial infections in humans, livestock, and fish. In 1962, as a by-product of antimalarial research, the first compound of this group emerged; nalidixic acid was found to be effective against some gram-negative microorganisms. In the 1970s, new 4-quinolones such as pipemidic acid, oxolinic acid, and cinoxacin were introduced, with marginal improvements over nalidixic acid.

Chemists have been able to synthesize thousands of quinolone derivatives, primarily with modifications at the N-1 position and at the C-6, C-7, and C-8 positions, with further variants on the horizon. The structural modifications incorporated into these new compounds gave improvements with regard to pharmacodynamic characteristics and pharmacokinetic profiles. In 1980s and 1990s, with the introduction of fluorine and piperazine substituents into the basic molecular structure, other important compounds such as norfloxacin, enrofloxacin, ciprofloxacin, and ofloxacin were developed as more potent quinolone compounds (Hooper and Wolfson 1993). A chronological development of quinolones is depicted in Table 1.

Currently, several fluoroquinolones (FQs) are available as antibacterial agents for the treatment of animals, poultry, and fish in many countries, mainly for the treatment of pulmonary, urinary, and digestive infections. Some drugs, such as enrofloxacin and sarafloxacin, were specially developed as veterinary medicines. Ciprofloxacin, norfloxacin, and ofloxacin are used in human medicine (Alder et al. 2001). Flumequine, norfloxacin, and ofloxacin are used both as human and animal drugs (Choma 2003). The major metabolite of enrofloxacin is its human counterpart, ciprofloxacin, one of the most widely used human antibiotics in the world. Because FQs are very related compounds, both structurally and in their mode of action, it is possible to consider the total concentration of FQs in the environment for their risk characterization. FQs are effective against a broad spectrum of pathogenic gram-negative and gram-positive bacteria and mycoplasmas (Hooper and Wolfson 1993).

Table 1. Chronological Development of Quinolones as Drugs.

Year	Name of drug	Developed by
1962	Nalidixic acid	Lappin
1966	Oxolinic acid	Warner-Lambert
1967	Piromidic acid	Dainippon
1970	Cinoxacin	E. Lilly
1972	Miloxacin, Rosoxacin	Sumitomo, Sterling
1973	Flumequine	Riker
1974	Pipedimic acid	Dainippon
1978	Norfloxacin	Kyorin
1979	Pefloxacin	Roger Ballan (Rhône-Poulenc)
1980	Enoxacin	Dainippon
1981	Fleroxacin	Kyorin
1982	Ofloxacin	Daiichi
1987	Ciprofloxacin	Bayer AG
1987	Enrofloxacin	Bayer AG
1985	Tosufloxacin	Toyama, Abbott
1987	Sparfloxacin	Dainippon
1989	Grepafloxacin	Warner Lambert
1991	Danofloxacin	Pfizer
1992	Trovafloxacin	Pfizer
1994	Levofloxacin	Daiichi
1994	Sarafloxacin	Abbott/Fort Dodge
1994	Orbifloxacin	Dainippon/Schering
1995	Marbofloxacin	Vetoquinol/Pfizer
1995	Moxifloxacin	Bayer AG
1996	Prulifloxacin	Nippon
1996	Difloxacin	Fort Dodge
2002	Ibafloxacin	Intervet

II. Estimated Usage

Internationally available data on the consumption of antibiotics are inadequate and contradictory (Kümmerer 2004). It is estimated that worldwide about 100,000–200,000t antibiotics are used (Wise 2002). In 1999, the total amount of antibiotics used in the European Union and Switzerland was 13,288t, of which 8,637t (65%), 3,854t (29%), and 797t (6%) were used in human medicine, veterinary medicine, and as growth promoters, respectively (FEDESA 2001). In the veterinary field, more than 70% of all consumed pharmaceuticals are antibiotic agents (Halling-Sørensen et al. 1998). In Italy, the theoretical environmental load in 2001 was reported as 2.96t/yr (Zuccato et al. 2004). Considering the United States, European Union (EU), Japan, and South Korea, quinolone production and usage is around 50t as proprietary products and 70t as generic quinolones; annual quinolone consumption for humans and animals in China is about 1,350 and 470 t, respectively (WHO 1998). The importance of FQs in the world market is gradually increasing. As of 1999, the global value of ciprofloxacin sales exceeded 1.3 billion dollars and the next most important FQ, ofloxacin, accounted for sales of approximately 900 million dollars (Nakata et al. 2005).

III. Occurrence

FQs may reach the environment through a variety of human and veterinary pathways, mostly via either human excretion into wastewaters or dispersion of manure onto agricultural soils. FQs are primarily excreted unchanged (Sörgel and Kinzig 1993). Sludge recovered from wastewater recycling activities may be applied directly to land as fertilizer. The presence of FQs has already been evidenced in topsoil samples where sewage sludge had been applied (Golet et al. 2003). Interestingly, sewage sludges from different waste treatment plants in Switzerland were reported to contain ciprofloxacin and norfloxacin in the range of 1.4–2.4 mg/kg dry matter (Alder et al. 2004). It is believed that a significant quantity of unchanged FQs may be introduced into the environment through wastewater from clinical settings (Hartmann et al. 1998). In wastewater from a hospital in Switzerland, the levels of ciprofloxacin and norfloxacin were 17.2–29.4 μg/L and 2.6–7.9 μg/L, respectively, over a 24-hr sampling period (Pham Thi 2003).

Other possible arenas for environmental contamination with antibiotics are either landfills from hospital/municipal waste or the disposal of wastes from pharmaceutical production sites. This is not a problem with modern landfills, which are equipped with protective barriers and leachate collection systems; however, potential risks exist with older landfills that do not possess the modern facilities. Although we are unable to find any such report on FQs, there are reports on other pharmaceuticals (Holm et al. 1995; Ahel and Jeličic 2001). Drug residues in the leachate may reach shallow groundwater and surface waters during the disposal of drug wastes to unlined landfill sites (Holm et al. 1995). Most animal waste reaches soil as manure and, as a result, the metabolized or unmetabolized animal drugs contained in liquid or solid animal waste may end up in the soil and subsequently be transported to surface water and groundwater (Hamscher et al. 2000, 2004). In intensive fish farming, infections are treated with antimicrobial agents, which are typically in the form of feed additives, and placed directly in the water; this leads to their direct entrance into the aquatic environment, causing a buildup of antibiotic residues in the water and sediments. Also, the buildup of antibiotics in fauna may not be ruled out. Residues of flumequine (0.06–1.12 μg/g) and oxolinic acid (0.08–15.74 μg/g) were reported in the nontarget surrounding wild fish population and other marine animals after the medication of cultivated fish (Ervik et al. 1994).

IV. Chemistry, Bioactivity, and Mode of Action

FQs are obtained by adding a fluorine group at C-6 and a piperazinyl (or piperazine derivative) group at C-7 to the core structure of quinolones (Table 2); in this way, the antibacterial activity has been greatly extended. Their bioactivity is mainly dependent on the presence of an aromatic fluorine substituent at C-6 (Wetzstein 2001). The carboxylic acid at position 3 and the ketone group at position 4 are necessary for DNA gyrase inhibition, whereas substitutions at positions 1 and 7 influence the potency and biological spectrum of activity of the drugs (Van Hoof et al. 2005). Quinolones have intrinsic gram-negative activity. Addition of a fluorine group increases the lipophilicity of the compounds and gives them greater gram-positive activity.

The active sites of FQs are illustrated in Fig. 1. Most recently, more attention has been paid to chiral chemistry of fluoroquinolones, as an improvement in antibacterial activity is observed when a chiral group is in close proximity to the quinolone core, such as at N-1 or C-7 (Bhanot et al. 2001). Levofloxacin, the S(–) enantiomer of ofloxacin, is now on the market. Ofloxacin is a tricyclic compound with a methyl group at the asymmetrical C-3 position in the oxazine ring. Levofloxacin is 8–28 times more potent than the R(+)-ofloxacin and twice as active as the racemic mixture (Bryskier and Chantot 1995). Generally, FQs are chemically stable and insensitive to hydrolysis and high temperatures but are susceptible to UV light (Thiele-Bruhn 2003). It is possible for them to chelate with cations, such as aluminium, magnesium, calcium, iron, and zinc. In acidic quinolones (without piperazinyl group), such as oxolinic acid, nalidixic acid, and flumequine, pK_a values for the carboxylic group range from 6.0 to 6.8 and from 5.7 to 6.3 in piperazinyl quinolones, while those for the protonated amino group are higher, i.e., 7.6–8.3 (Barbosa et al. 2001).

The site and mode of action of different antibacterial agents are different (Fig. 2). FQs interfere with bacterial DNA metabolism by inhibiting the enzymes topoisomerase II (DNA gyrase) and topoisomerase IV, which are important for long bacterial DNA to fit into the cell through supercoiling. FQs predominantly inhibit DNA gyrase in gram-negative bacteria and topoisomerase IV in gram-positive bacteria. During the supercoiling process, DNA gyrase cleaves both DNA strands and forms a "quinolone-binding pocket." In a dimer structure, two quinolone molecules are bound to the pocket electrostatically. It is believed that the amine substituents at C-7 also participate to strengthen the attachment to give the drug–enzyme–DNA complex. Thus, the supercoiling process is blocked, facilitating the synthesis of a repair enzyme (exonuclease), which results in an uncoordinated repair process, irreversible damage to the DNA, and, finally, death of the cell (Hooper and Wolfson 1993; Morais-Cabral et al. 1997). The toxicity of FQs to mammalian cells is in general low because the enzyme analogue to bacterial DNA gyrase in eukaryotes is 100- to 1,000-fold less susceptible to gyrase

Table 2. Nomenclature, Chemical Structure, and Use of Some Common Fluoroquinolones.

Name	Nomenclature	Structure	Use
Ciprofloxacin CAS No.: 86483-48-9 Chemical Formula: $C_{17}H_{18}FN_3O_3$ Mol. Wt.: 331.35	1-Cyclopropyl-6-fluoro-1,4-dihydro-4-oxo-7-(1-piperazinyl)-3-quinoline carboxylic acid	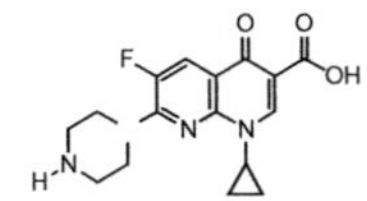	Human and veterinary use
Danofloxacin CAS No.: 119478-55-6 Chemical Formula: $C_{19}H_{20}FN_3O_3$ Mol. Wt.: 453.49	7-[(1′s,4′s)-5′-Methyl-2′,5′-diazabicyclo[2.2.1]hept-2′-yl]-1-cyclopropyl-6-fluoro-1,4-dihydro-4-oxo-3-quinoline carboxylic acid	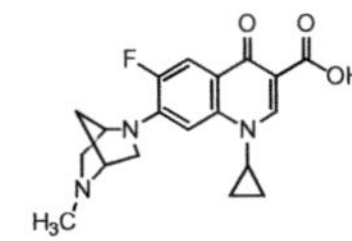	Veterinary use
Difloxacin CAS No.: 98106-17-3 Chemical Formula: $C_{21}H_{19}F_2N_3O_3$ Mol. Wt.: 399.37	6-Fluoro-1-(4-fluophenyl)-1,4-dihydro-7-(4-methyl-1-piperazinyl)-4-oxo-3-quinoline carboxylic acid	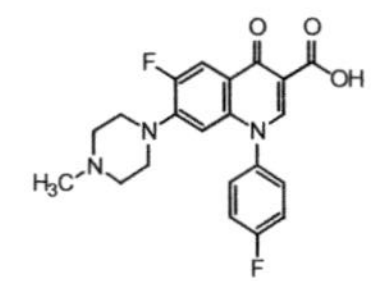	Veterinary use
Enoxacin CAS No.: 74011-58-8 Chemical Formula: $C_{15}H_{17}FN_4O_3$ Mol. Wt.: 320.33	1-Ethyl-6-fluoro-1,4-dihydro-4-oxo-7-(1-piperazinyl)-1,8-naphthyridine-3-carboxylic acid	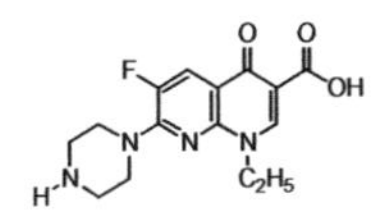	Human use
Enrofloxacin CAS No.: 93106-60-6 Chemical Formula: $C_{19}H_{22}FN_3O_3$ Mol. Wt.: 359.40	1-Cyclopropyl-6-fluoro-1,4-dihydro-4-oxo-7-(4-ethyl-1-piperazinyl)-3-quinoline carboxylic acid	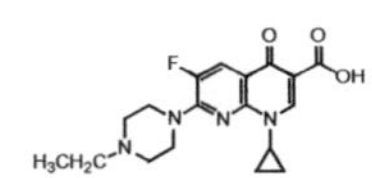	Veterinary use

Fleroxacin CAS No.: 79660-72-3 Chemical Formula: $C_{17}H_{18}F_3N_3O_3$ Mol. Wt.: 369.34	6,8-Difluoro-1-(2-fluoroethyl)-1,4-dihydro-7(4-methyl-1-piperazinyl)-4-oxo-3-quinoline carboxylic acid	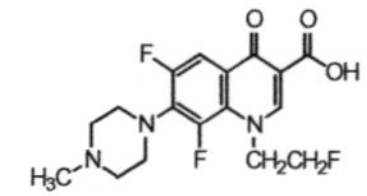	Human use
Flumequine CAS No.: 42835-25-6 Chemical Formula: $C_{14}H_{12}FNO_3$ Mol. Wt.: 261.25	9-Fluoro-6,7-dihydro-5-methyl-1-oxo-1H, 5H-benzo-(IJ)-quinolizine-2-carboxylic acid	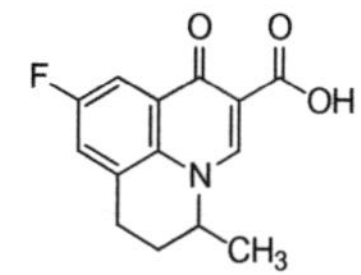	Veterinary use
Gatifloxacin CAS No.: 112811-59-3 Chemical Formula: $C_{19}H_{22}FN_3O_4$ Mol. Wt.: 375.39	(±)-1-Cyclopropyl-6-fluoro-1,4-dihydro-8-methoxy-7-(3-methyl-1-piperazinyl)-4-oxo-3-quinoline carboxylic acid	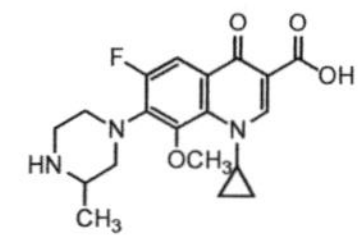	Human use
Grepafloxacin CAS No.: 119914-60-2 Chemical Formula: $C_{19}H_{22}FN_3O_3$ Mol. Wt.: 359.39	1-Cyclopropyl-6-fluoro-1,4-dihydro-5-methyl-7-(3-methyl-1-piperazinyl)-4-oxoquinoline-3-carboxylic acid	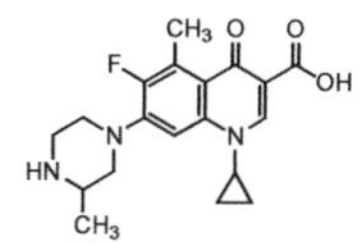	Human use
Levofloxacin CAS No.: 100986-85-4 Chemical Formula: $C_{18}H_{20}FN_3O_4$ Mol. Wt.: 361.37	*S*-isomer of racemic ofloxacin	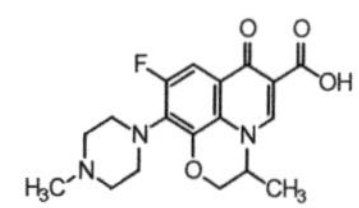	Human use

Table 2. *Continued*

Name	Nomenclature	Structure	Use
Lomefloxacin CAS No.: 98079-52-8 Chemical Formula: $C_{17}H_{19}F_2N_3O_3$ Mol. Wt.: 351.35	1-Ethyl-6,8-difluoro-1,4-dihydro-7-(3-methyl-1-piperazinyl)-4-oxo-3-quinoline carboxylic acid	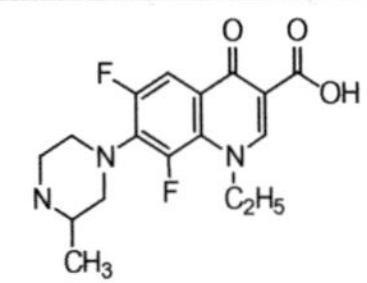	Human use
Marbofloxacin CAS No.: 115550-35-1 Chemical Formula: $C_{17}H_{19}FN_4O_4$ Mol. Wt.: 362.36	9-Fluoro-2,3-dihydro-3-methyl-10-(4-methyl-1-piperazinyl)-7-oxo-7H-pyrido[3.2.1-ij][4.2.1] benzoadiazine-6-carboxylic acid	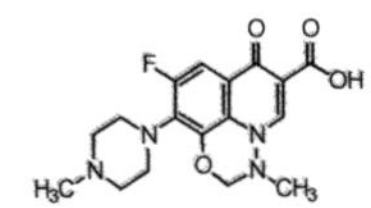	Veterinary use
Moxifloxacin CAS No.: not found Chemical Formula: $C_{21}H_{24}FN_3O_4$ Mol. Wt.: 401.43	1-Cyclopropyl-7-(2,8-diazobicyclo[4.3.0]nonane)-6-fluoro-8-methoxy-1,4-dihydro-4-oxo-3-quinoline carboxylic acid	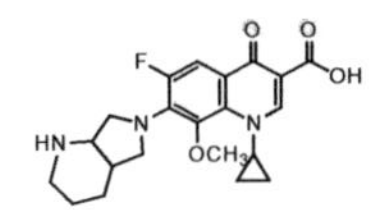	Human use
Norfloxacin CAS No.: 70458-96-7 Chemical Formula: $C_{16}H_{18}FN_3O_3$ Mol. Wt.: 319.33	1-Ethyl-6-fluoro-1,4-dihydro-4-oxo-7-(1-piperazinyl)-3quinoline carboxylic acid	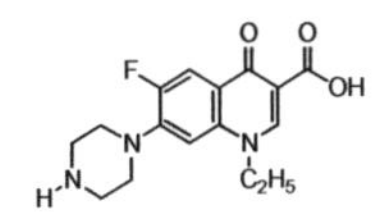	Human and veterinary use

Ofloxacin CAS No.: 83380-47-6 Chemical Formula: $C_{18}H_{20}FN_3O_4$ Mol. Wt.: 361.37	(±)-9-Fluoro-2,3-dihydro-3-methyl-10-(4-methyl-1-piperazinyl)-7-oxo-7 *H*-pyrido [1,2,3-*de*]-1,4-benzoxazine-6-carboxylic acid	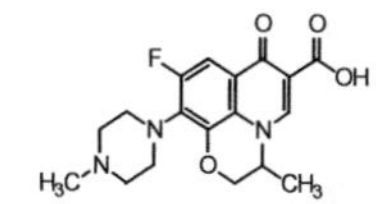	Human and veterinary use
Pefloxacin CAS No.: 70458-92-3 Chemical Formula: $C_{17}H_{20}FN_3O_3$ Mol. Wt.: 333.36	1-Ethyl-6-fluoro-1,4-dihydro-7-(4-methylpiperazin-1-yl)-4-oxoquinoline-3-carboxylic acid	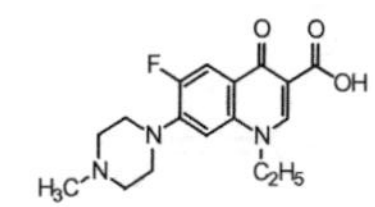	Human use
Sarafloxacin CAS No.: 98105-99-8 Chemical Formula: $C_{20}H_{17}F_2N_3O_3$ Mol. Wt.: 385.37	6-Fluoro-1-(4-fluorophenyl)-7-(3-methyl-1-piperazinyl)-1,4-dihydro-4-oxo-3-quinoline carboxylic acid	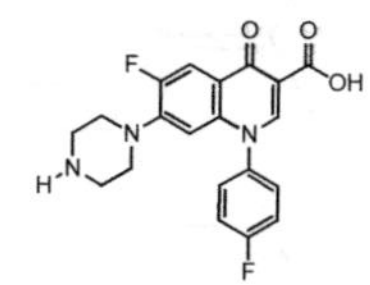	Veterinary use
Tosufloxac.in CAS No.: 108138-46-1 Chemical Formula: $C_{19}H_{15}F_3N_4O_3$ Mol. Wt.: 404.35	(±)-7-(3-Amino-1-pyrolidinyl)-1-(2,4-difluorophenyl)-6-fluoro-1,4-dihydro-4-oxo-1,8-naphthyridine-3-carboxylic acid	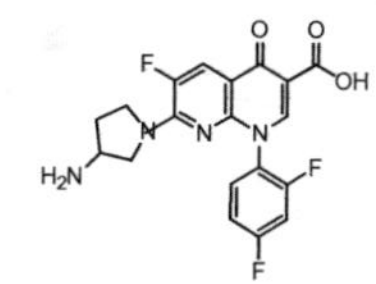	Human use

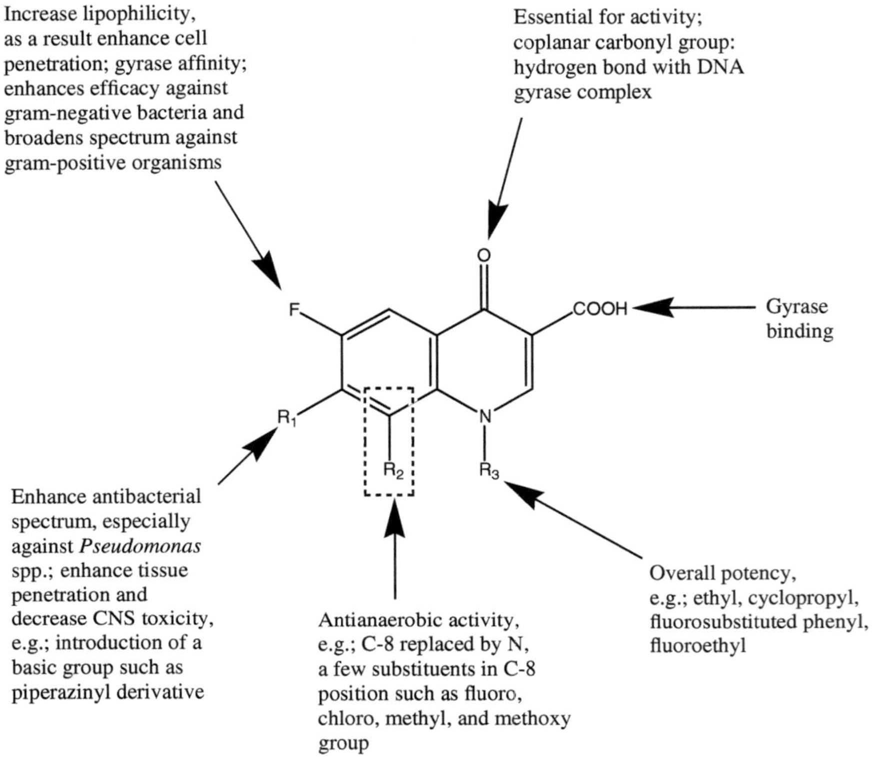

Fig. 1. Active sites of fluoroquinolones.

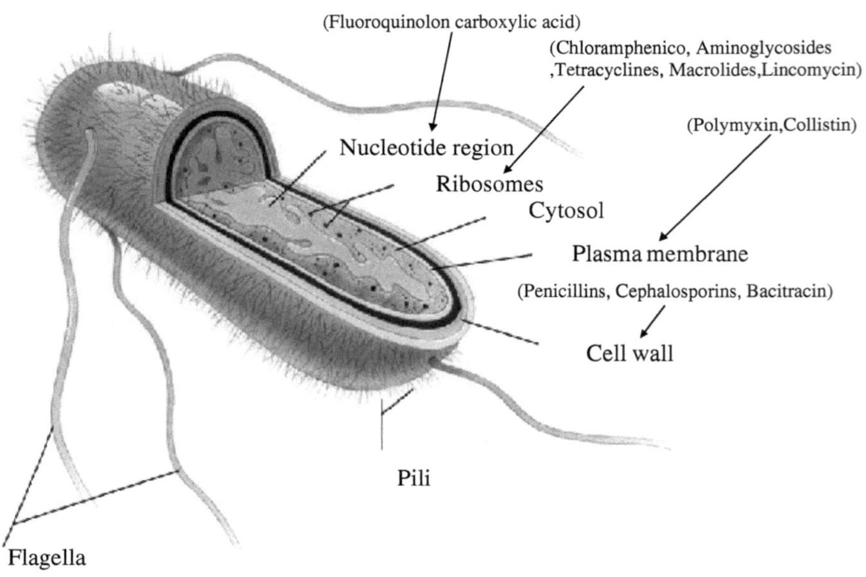

Fig. 2. Sites of action of antibacterial drugs in bacterial cell.

inhibitors (Booth 1994); this explains their broad safety margin in human and animal therapy.

V. Degradation

Degradation, both biotic and abiotic, is one of the significant and relevant elimination processes for antibiotics in soil, water, and other matrices. This process counteracts the accumulation and adverse biological effect of the drugs in the environment, as well as their transport to surface waters or groundwaters. However, the exhibition of antimicrobial activity by many FQ metabolites cannot be ruled out (Marengo et al. 1997; Wetzstein et al. 2000). Under anaerobic conditions, FQs show lower degradation than under aerobic conditions, e.g., as shown for ciprofloxacin (Ingerslev et al. 2001; Halling-Sørensen et al. 2003). Being antimicrobial, it is understood that FQs are not very susceptible to microbial degradation (Al-Ahmad et al. 1999; Kümmerer et al. 2000; Marengo et al. 2001). However, ciprofloxacin and enrofloxacin showed notable degradation by hydroxyl radical-mediated enzyme systems characteristic of brown-rot fungi (Parshikov et al. 1999; Wetzstein et al. 1999). In the presence of wood-rotting fungi, metabolism of FQs at the amine position is seen without degradation of the heterocyclic ring (Wetzstein 2001). The high reactivity and fast reactions of FQs with manganese oxide have been demonstrated (Zhang and Huang 2005). These results demonstrate the influence of manganese oxide, which is present in soils as a naturally occurring reactant, on the abiotic degradation of FQs in the environment. It has been shown that the piperazine moiety of FQs is the predominant adsorptive and oxidative site and undergoes dealkylation and hydroxylation, with the quinolone ring remaining essentially intact. Therefore, due to the high affinity of FQs to soil and sediments (Tolls 2001), reactions of FQs with MnO_2 may play a governing role to influence their fate in soil and water systems. Moreover, the dealkylated oxidation products of FQs would lead to a reduction in antimicrobial activity (Zhang and Huang 2005). It is possible that the persistence of drugs in soil and manure may vary with the temperature and with the types of manure and compounds (Van Dijk and Keukens 2000; Boxall et al. 2003).

A. Biotransformation

The biotransformation of drugs by different organisms may follow different pathways. In general, antibiotics administered in animals undergo phase 1 reactions, which include oxidation, reduction, and hydrolysis, followed by phase II reactions to form readily excreted water-soluble polar metabolites (Daughton and Ternes 1999; Tolls 2001). Conjugation of the parent compound or its metabolite containing —OH, —COOH, —NH_2, etc., functional groups with D–glucuronic acid is considered to be the most common phase

II reaction for antibiotics. The degree of metabolism of FQs in animals is considered to be moderate to high [moderate, 20%–80%; high, >80% metabolism (Boxall et al. 2002)]. In animals, there is a general trend for FQs to be attacked at the carboxyl group and at the piperazine or the *N*-methyl piperazine ring (Borner et al. 1990). *N*-oxidation of the methylpiperazine ring is possible in ofloxacin, pefloxacin, and fleroxacin (Sudo et al. 1986; Lode et al. 1987; Nakashima et al. 1988). Nitrogen acetylation may also be possible by the removal of two ethylene carbons from the piperazine ring and by other reaction, as evidenced for enoxacin (Nakamura et al. 1983). The major mammalian metabolites obtained from ciprofloxacin show less antibacterial activity than the parent compound (Zeiler et al. 1987). Enrofloxacin may be transformed to ciprofloxacin in animals by *N*-dealkylation of the ethylpiperazine ring (Tyczkowska et al. 1989). This reaction is governed by the monooxygenase-mediated α-hydroxylation of the ethyl group via an unstable hemi-aminal intermediate (Azerad 1999). It is presumed that the conversion of enrofloxacin to ciprofloxacin might also occur in the fungus *Mucor ramannianus* (Parshikov et al. 2000). In this study, the authors were able to identify *N*-acetylciprofloxacin but not ciprofloxacin. It was proposed that ciprofloxacin was not detected because of the fast acetylation step. In addition to *N*-acetylciprofloxacin, they identified two other metabolites, namely enrofloxacin *N*-oxide and desethylene-enrofloxacin. In another study, ciprofloxacin was found to be metabolized to *N*-acetylciprofloxacin by *Mucor ramannianus* (Parshikov et al. 1999).

Hydroxylation, decarboxylation, defluorination, and removal of part of or the whole piperazine ring were found to be significant reactions during the biotransformation of enrofloxacin and ciprofloxacin in basidiomycetes fungi (Wetzstein et al. 1997, 1999). The widespread potential for ciprofloxacin degradation among basidiomycetes inhabiting various environments, including agricultural soils and animal dung, has been established, and 11 metabolites were identified. These are grouped into four categories: monohydroxylated congeners, dihydroxylated congeners, an isatin-type compound (elimination of C-2), and metabolites indicating both elimination and degradation of the piperazinyl moiety (Wetzstein et al. 1999). In humans, ciprofloxacin may be metabolized by sulfation and by oxidation of its piperazine moiety. Glucuronidation has also been found in various animal species (Dalhoff and Bergan 1998). Demethylation of danofloxacin at the *N*-methylpiperazine ring by *Rhizopus arrhizus* has also been observed (Chen et al. 1997). The presence of zygomycetous fungi such as *Rhizopus arrhizus* and *Mucor ramannianus* in soil and decomposing organic matter (Griffin 1972) indicates the inactivation potential for FQs in nature and also points out their ecological significance. Moreover, their biotransformed products are reported to be less toxic than the parent compound (Wetzstein et al. 1997). Norfloxacin was reported to produce 4 metabolites, 7-amino-1-ethyl-6-fluoro-4-oxo-1,4-dihydroquinolone-3-carboxylic acid, *N*-formylnorfloxacin, *N*-acetylnorfloxacin, and desethylene-*N*-acetylnorfloxacin, by the action of a nonpathogenic fungus, *Pestalotiopsis guepini*, in poultry litter (Williams et al. 2004).

B. Photodegradation

Although the most important route of attenuation of xenobiotics in the environment remains their biodegradation, photodegradation may also be an important elimination process. Because photodegradation is likely to occur on the top layer of the soil surface, photolysis of compounds depends on agricultural practices, such as the timing and depth of ploughing, and also on the plant canopy. However, photolysis of compounds in aqueous systems is of great significance. FQs are susceptible to photodegradation in aqueous media, and direct UV photolysis or radical-mediated photolysis is a common phenomenon for FQs (Fasani et al. 1999, 2001; Mella et al. 2001). A half-life <1 hr was determined when sarafloxacin was subjected to photodegradation in water (Davis et al. 1993). Flumequine was 96% photodegraded in water after 9 d (Lunestad et al. 1995). Phototransformation is reported to be the main attenuation mechanism for ciprofloxacin and ofloxacin in shallow rivers and lakes, with half-lives of approximately 30 min near the surface of sunlit surface waters (Golet et al. 2002a,b).

A detailed study on the photochemically induced decomposition of fluoroquinolone carboxylic acids at concentrations of 10 mg/L in pure water and an irradiation intensity of 200 W/m^2 (xenon lamp) showed half-lives of 20.6 min (danofloxacin), 36.2 min (enrofloxacin), 90.2 min (ciprofloxacin), and 105.9 min (norfloxacin) (Burhenne et al. 1997a). Depending on the season and degree of latitude, the environmental half-life of enrofloxacin was found to be 1.8–55.4 hr. Photolytic pathways for FQs are depicted in Fig. 3. Similar to its microbial transformation (Parshikov et al. 2000), enrofloxacin is phototransformed to ciprofloxacin by deethylation of the ethylpiperazine ring. Oxidation, dealkylation, and cleavage of the piperazine ring are the major reaction mechanisms in the photodecomposition of FQs. Polar photometabolites, such as pyridine dicarboxylic and tricarboxylic acids, are also expected during photodegradation of FQs in aqueous solution (Burhenne et al. 1997b). It has also been established that photomineralization is a common phenomenon for FQs. The pH, presence of cosolutes, etc., greatly influence the photolysis of FQs, with respect to both the kinetics and the nature of the photoproducts (Fasani et al. 1999, 2001). Ofloxacin photodegrades more slowly than ciprofloxacin, enrofloxacin, and norfloxacin, undergoing defluorination while the other three opt for dealkylation (Fasani et al. 1999). This difference is perhaps due to the presence of a electron-withdrawing substituent at position-5 on the aromatic ring of ofloxacin, while the other three possess a hydrogen at this position.

Interestingly, antimicrobial activity was found in the photodegradation products of ofloxacin and levofloxacin (Sunderland et al. 1999), but no

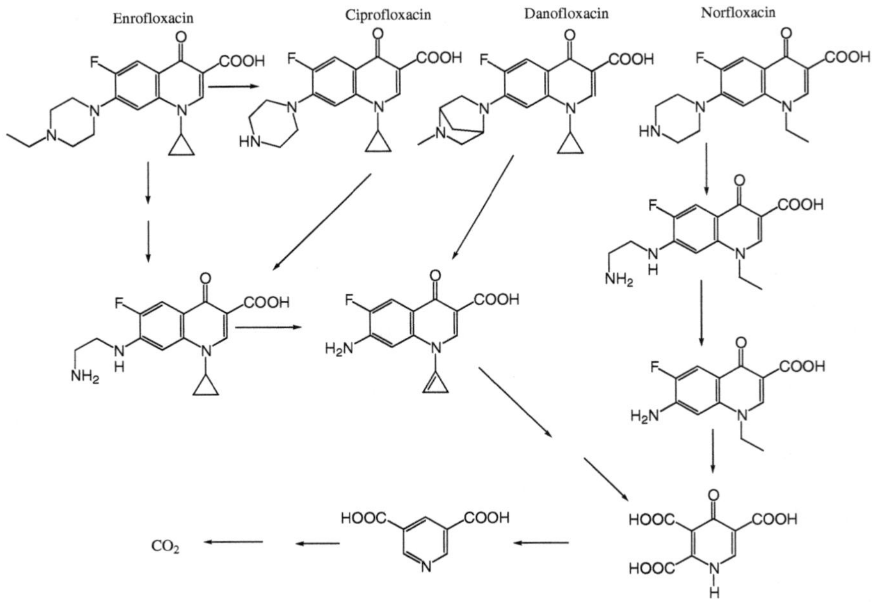

Fig. 3. Proposed scheme for photolysis of fluoroquinolones.

detectable antimicrobial activity was reported for the photodegradation products of ciprofloxacin (Philips et al. 1990; Sunderland et al. 2001). Thus, it may be postulated that the dealkylated products have much lower antibacterial activity than the defluorinated products (Zhang and Huang 2005). Recently, rapid photodegradation of ciprofloxacin and lomefloxacin, to yield photolysis products of lower biological activity than the parent compounds, has been reported (Matzen et al. 2005). It is proposed that both drugs degrade by direct photolysis. Sitafloxacin undergoes photodegradation in aqueous solution through dechlorination. (Araki et al. 2002). The rate of photolysis of sitafloxacin was higher in neutral solutions than the rates observed in acidic and alkaline solutions.

VI. Persistence

A. Soil and Sediments

Sarafloxacin in marine sediments exhibited very little degradation, with the degradation amounting to 0.06% after 83 d (Marengo et al. 1997). Half-lives of 60 and >300 d were observed for flumequine at 0–1 cm and 5–7 cm depths in marine sediment, respectively (Hektoen et al. 1995). In field soils of different types, more than 80% degradation was observed for sarafloxacin after 80 d (Marengo et al. 1997). For danofloxacin, a half-life of 87–143 d was observed in three different soils (Chen et al. 1997). The persistence of FQs in sludge-treated soils for several months after application has also been reported (Golet et al. 2002b). Thus, it may be concluded that FQs are rather persistent in both field studies and laboratory tests. The strong binding of FQs to soil and sediment components delays their biodegradation and explains the recalcitrance of FQs (Hektoen et al. 1995; Marengo et al. 1997). More critically, unspecified adsorption to dissolved organic matter causes a decline in the degradation, e.g., as in the case of enrofloxacin (Wetzstein et al. 1997).

B. Surface Water, Sewage Sludge, and Wastewater Effluent

In general, biodegradation in surface waters is considered to be slower than in sewage systems because of lower density and diversity of bacteria. In a test simulating surface water, ciprofloxacin was found to be stable (Al-Ahmad et al. 1999). FQs have been reported in the Glatt River, Switzerland, at concentrations to 0.02 μg/L (Golet et al. 2002a), which might be due to receiving surface water from a wastewater treatment plant (WWTP) representing a densely populated region. An attempt has been made to obtain the concentration of FQs in surface water from their measured environmental concentration (MEC) in tertiary wastewater effluents after wastewater treatment (Golet et al. 2001). Assuming a dilution factor of 10 from the MEC, the concentration of ciprofloxacin and norfloxacin was estimated to be about 5–12 ng/L. Interestingly, these values are one to two orders of magnitude lower than the predicted environmental concentration (PEC) estimated by conventional risk assessments (Kümmerer et al. 2000; Halling-Sørensen et al. 2000).

In the U.S., four FQs (ciprofloxacin, norfloxacin, enrofloxacin, and sarafloxacin) were reported with mean concentrations from nondetectable to 0.12 μg/L in surface waters (Kolpin et al. 2002). However, in another study, no FQs were detected in 189 water samples collected from 13 U.S. fish hatcheries, although 14% of the samples were reported to contain tetracycline, oxytetracycline, and sulfadimethoxine (Thurman et al. 2002). Untreated groundwater used for drinking could be a source for the spread and transfer of antibiotic-resistant strains (Sørum and L'Abée-Lund 2002). However, FQs are reported to sorb strongly at topsoil, preventing their leaching to lower layers (Nowara et al. 1997), and thus do not pose a real threat to groundwater resources. In Germany, FQs (ciprofloxacin and ofloxacin) occurred sporadically in surface water samples at concentrations <10 ng/L (Christian et al. 2003).

FQ concentration in raw sewage and final wastewater effluents from a densely populated region in Switzerland ranged 255–568 ng/L and 36–106 ng/L, respectively (Golet et al. 2002a), which clearly shows that wastewater treatment significantly reduces the concentration of FQs and emphasizes the importance of sludge treatment strategies. The level of removal of FQs from the water stream during wastewater treatment was between 79% and 87%. An additional decrease in FQ concentration (66% for ciprofloxacin and 48% for norfloxacin) was also observed during their transport in the river from the wastewater treatment but, because complete removal was not achieved, residual amounts of ciprofloxacin and norfloxacin were emitted into neighboring waters. It is important to note that although the ultimate concentration of FQs reaching the river was low, the trace amounts emitted into the receiving waters were persistent. As a result, the low biodegradability of quinolones (Al-Ahmad et al. 1999) may influence the aquatic environment, because a few FQs, e.g., ofloxacin, have been found to be genotoxic (Kümmerer et al. 2000), and the presence of FQs may have negative impact on self-cleaning capacity of surface water because of the low bacterial density present. However, wastewater treatment may be considered as an important and efficient elimination step for FQs before they enter rivers. No veterinary-use FQs, such as enrofloxacin, danofloxacin, or difloxacin, were detected in urban wastewaters (Golet et al. 2001), which establishes the great influence of human use on the occurrence of FQs in river waters.

Ciprofloxacin (max., 0.12 mg/L), norfloxacin (max., 0.11 μg/L), and ofloxacin (max., 0.10 μg/L) were frequently detected in the final effluent of eight wastewater treatment plants located in five Canadian cities (Miao et al. 2004). In the U.S., ciprofloxacin and ofloxacin were repeatedly detected in municipal wastewaters at 80 ng/L to 2 μg/L (Renew and Huang 2004). Ofloxacin was reported to be present in the secondary and final effluents of a WWTP in East Lansing, MI, at concentrations of 204 and 100 ng/L, respectively, and, based on the mass flow calculation, a discharge of ofloxacin to the river at a concentration of 4.8 g/d was expected (Nakata et al. 2005). In the same study, the authors reported the presence of ciprofloxacin, norfloxacin, and lomefloxacin at concentrations <45 ng/L in wastewater effluents. High concentrations of ofloxacin have also been found in the effluents from sewage treatment plants in European countries [France, 330–510 ng/L; Italy, 290–580 ng/L; Greece, 460 ng/L (Andreozzi et al. 2003)]. Ciprofloxacin and norfloxacin have been detected in the effluents of WWTPs in Switzerland (Golet et al. 2002a,b) and several European countries (Andreozzi et al. 2003). In Germany, the presence of 0.7–124.5 μg/L ciprofloxacin has been reported in hospital wastewaters (Hartmann et al. 1999).

VII. Sorption

The dissipation of a compound in a test system does not necessarily mean its biotic or abiotic degradation. The compound may have a tendency to bind to soil particles or to interact with ions present in the system (Hektoen et al. 1995; Marengo et al. 1997; Thiele-Bruhn 2003). Oxolinic acid, flumequine, and sarafloxacine have been reported to adsorb to sediment of marine origin and found to be very persistent (Hektoen et al. 1995), with half-lives >300 d being estimated. In spite of binding to sediments, oxolinic acid exerted antibacterial activity for a long duration. However, in another experiment, a contradictory result occurred, whereby oxolinic acid exhibited no observable antibacterial activity in fish farm sediment after 10 d drug addition (Björklund et al. 1991). The exhibition of biological activity of FQs after their sorption to sediments may be influenced by the sediment composition, which determines the degree and strength of sorption. In a sewage treatment plant (STP) simulation, ciprofloxacin was found to be fairly well adsorbed, with about 65% eliminated by sorption and 30% detected in the effluent (Kümmerer et al. 2000). It is also established that wastewater treatment causes an 88%–92% reduction in the mass flow of FQs, which is primarily caused by sorption on sewage sludge (Xifra 2000; Golet et al. 2003). In another study, a strong sorption behavior of FQs, with special reference to ciprofloxacin and norfloxacin, was also established (Golet et al. 2002a,b). Therefore, the attenuation of these compounds during the wastewater treatment process was caused not by their degradation but by their sorption to the particulate matters; in this way, they reach the soil sorbed to sewage sludge, where they may persist for years. Although application of sewage sludge in agricultural fields has been forbidden in Switzerland since 2003, there are many countries where the practice of using sewage sludge in crop fields still exists, with the idea of its disposal and also of the sustainable principle of nutrient recycling in soil. Therefore, a clear knowledge on the behavior of FQs in sludge-treated soils is much needed.

FQ adsorption appeared to be positively correlated with the organic carbon content of the soil. In particular, ofloxacin so strongly adsorbed to soil rich in organic carbon that, in a lysimetric study, there was no recovery of ofloxacin in the leachate, indicating a much lower mobility in the soil for this compound (Stamatelatou et al. 2003). However, it is also important to note that the sorption of antibiotics to solid particles reduces both its concentration in solution and its bioavailability, thus reducing the photodegradation and biodegradation potential of FQs (Sithole and Guy 1987; Lützhøft et al. 2000). It is interesting to note that strong sorption of FQs on the topsoil reduces the threat of surface water and groundwater contamination, but that at the same time the availability of these compounds to soil-dwelling organisms becomes relevant. However, the high values of $k_{d,\ DOM}$ (DOM, dissolved organic matter) for quinolone carboxylic acids indicate that DOM may also compete with soil solids for quinolone carboxylic acid molecules (Lützhøft et al. 2000) and association of FQs with DOM may increase their presence in soil water, causing DOM-facilitated transport in the soil. The FQs may form complexes with multivalent cations, and the stability of the Ca and Mg complexes is significantly lower than for the Al and Fe complexes. The high stability constants of FQs with Al(III) and Fe(III) imply that FQs may undergo complex formation at aluminum and iron hydroxide surfaces

(Tolls 2001), which are in their amorphous states, coating the soil solids and the edges of clay minerals and thus facilitating sorption (Sithole and Guy 1987). FQs have been reported to be stongly bound by human feces (Van Saene et al. 1986; Edlund et al. 1988) and soil (Velagaleti et al. 1993; Marengo et al. 1997; Nowara et al. 1997). Deprotonated FQ-carboxylic acids interact with exchangeable cations bound to the negatively charged mineral surfaces, and the binding of FQs to clay minerals, especially to montmorillonite, is positively correlated to the expansion of interlayer spacing, because the adsorption of FQs at the clay mineral montmorillonite occurred between the layers (Nowara et al. 1997). Sorption of FQs is favored by their plane structure.

VIII. Analytical Methods

In this review, an attempt is made to report recent advances in analytical techniques used for the determination of FQs in water and biological samples. In water samples, FQs may be concentrated by solid-phase extraction (SPE) using mixed-phase cation-exchange (MPC) disk cartridges followed by elution with 5% methanolic ammonia (Golet et al. 2001; Nakata et al. 2005). Ultrapure water extracts from bovine muscle/aquacultured products were reportedly cleaned up using an Isolute C_{18} SPE cartridge followed by elution with 1% trifluoroacetic acid in acetonitrile (Van Hoof et al. 2005). This clean-up process is also applicable for milk, which should be deproteinized by precipitation using 20% trichloroacetic acid in methanol. Purification of edible animal tissue samples from endogenous interferences may also be performed using Oasis HLB cartridges (Samanidou et al. 2005a). Occasionally, tissue sample preparations have been carried out by adding phosphate buffer (pH 7.4, 0.1 M), followed by extraction with trichloromethane (Garcia et al. 2005).

Solvent extraction methods in biological samples sometimes create problems by the formation of emulsions and foams. Attempts are now being made to use supercritical fluid extraction (SFE) as an efficient and alternative extraction method for FQs. SFE of FQs from chicken breast muscle showed a recovery value ranging from 101% to 104% (enrofloxacin; Shim et al. 2003) and from 70% to 87% (norfloxacin and ofloxacin; Shen et al. 2004). Methods used for FQs analysis in biological samples or in their pharmaceutical dosage forms are mostly based on liquid chromatography with ultraviolet (Marazuela and Moreno-Bondi 2004; Vílchez et al. 2004; Samanidou et al. 2005a,b), fluorescence (Shen et al. 2004; Marazuela and Moreno-Bondi 2004; Garcia et al. 2005) or mass spectrometric detection (Schneider and Donoghue 2003; Nakata et al. 2005; Van Hoof et al. 2005). Liquid chromatography with mass detection is mostly chosen because it is sensitive, selective, and the preferred technique for confirmation of suspect residues, allowing multiresidue determination (Schneider and Donoghue 2003).

Various types of stationary phase (reversed-phase, polymer or phenyl columns) and mobile phase (changes in ionic strength, acidity, and/or the presence of modifiers such as citric acid, perchloric acid, or tertiary amines) have been used. An attempt was made to determine enrofloxacin and its metabolite ciprofloxacin in goat milk by LC-UV detection, combined with LC-MS for confirmation (Cinquina et al. 2003). However, the limit of quantification (LOC) for both analytes was 20 ng/mL. Later, a more sensitive method was described using LC with an FLD and UV-DAD detection system for multiresidue determination of FQs in milk (Marazuela and Moreno-Bondi 2004). With this method, six FQs were separated on a polar end-capped column (AQUA C_{18}), protected by a RP 18 guard column, within 13 min. A gradient program was used, with the mobile phase combining orthophosphoric acid and acetonitrile. Simultaneous determination of six quinolones in pig muscles at the 7.5 μg/kg level by liquid chromatography-atmospheric pressure chemical ionization mass spectrometry on a C_{18} column with a gradient elution was performed after extracting with phosphate buffer (pH 7.4) and purification on a C_{18} solid-phase extraction cartridge (Delepine et al. 1998).

A multiresidue method for 13 quinolones in feeds using photodiode-array and fluorescence detection in a liquid chromatographic system was studied and separation on a C_5 LUNA column was obtained in less than 27 min (Percorelli et al. 2003). In their pharmaceutical dosage forms, spectrofluorimetric methods could be applied successfully, with good precision and accuracy, for a few FQs (norfloxacin, ofloxacin, pefloxacin levofloxacin, lomefloxacin), through charge transfer complex formation with tetracyanoethylene (Du et al. 2004a) or with tetracyanoquinodimethane (Du et al. 2004b). Besides these techniques, gas chromatography (Pfenning et al. 1996; Asami et al. 2000), high-performance thin-layer chromatography (Wang et al. 2001; Choma et al. 2002), and capillary electrophoresis (Hernández et al. 2002; Barrón et al. 2002, 2003) have also been successfully used.

FQs have one or two chiral centers in their chemical structure and are available as racemates (ofloxacin, gemifloxacin, clinafloxacin), diastereoisomers (sparfloxacin), or pure enantiomers (levofloxacin, moxifloxacin). Concern over drug stereochemistry has increased recently due to the existing pharmacokinetic differences between the enantiomers of chiral drugs. As a result, concern over the methodology of their determination has also increased. HPLC separation of FQ sterioisomers has been extensively reviewed (Grellet et al. 2002) and is not considered here.

IX. Resistance

Of many alternatives, the food-borne route is considered to be one of the most important paths of resistance transmission from nonhuman sources (food-producing animals) to humans. Resistant bacteria emerging from food animals cause human infections and can also transmit their resistant determinants to human pathogenic bacteria in humans through horizontal transmission. In many cases, the use of antibiotics on farms has been found to be closely associated with the antimicrobial resistance in *Salmonella* isolated from humans (FAO/OIE/WHO 2003). An outbreak of human nalidixic acid-resistant *Salmonella typhimurium* DT 104 infections was reported at a pig farm in Denmark (Molbak et al. 1999) and at a dairy farm in the United Kingdom (Walker et al. 2000). Because of human dependency on FQs and the gradually increasing resistance of *Salmonella* and *Campylobacter* to FQs, the WHO consultation in the expert meeting in Geneva focused on the human health risks associated with the use of FQs in food animals and pointed out a possible correlation between the amount of antimicrobial agents used in food animals and antimicrobial resistance in selected bacteria (FAO/OIE/WHO 2003).

Occurrence of FQ-resistant *Campylobacter* in infected persons who were likely to have eaten chicken or turkey but not in healthy control subjects confirmed that domestically produced poultry is a source of domestically acquired FQ-resistant *Campylobacter* infections in the U.S. because chicken and turkey are not imported into the U.S. (Kassenborg et al. 2004). The emergence of FQ resistance in *Salmonella choleraesuis* following the use of FQs in pigs was also reported in Taiwan (Chiu et al. 2002). It has been suggested that risk related to toxicity and flora perturbation caused by residues of antimicrobials in foods is very low, but that the risk related to the development of antimicrobial resistance as a whole in bacteria may be of significant importance. It is believed that the ingestion of residues of antibiotics in food of animal origin poses a threat to human health by colonization barrier disruption, leading to pathogenic bacteria overgrowth or by exerting a selective pressure on the intestinal microflora favoring the growth of microorganisms with intrinsic or acquired resistance (FAO/OIE/WHO 2003). However, because of their strong binding to soil, FQs become nonbioavailable and are unlikely to exert a significant selection pressure (Van Saene et al. 1986; Velagaleti et al. 1993). At subinhibitory concentrations, antibiotics may have a large influence on cell functions and may alter the expression of virulence factors and thus the transfer of antibiotic resistance (Ohlsen et al. 1998). The main mechanism of resistance to FQs occurs by chromosomal mutations in the genes encoding for DNA gyrase and topoisomerase IV, decreasing the affinity of the drugs for the binding sites on the enzymes and making the drugs less effective inhibitors.

X. Risk Assessment

The widespread release of active molecules suggests a potential risk for the environment and the equilibrium of the ecosystem. The predicted environmental concentration (PEC) in different environmental compartments is an important and widely accepted decision parameter in the risk assessment procedure for antibiotics. The PEC value should be compared with predicted no effect concentration (PNEC). Values of PEC/PNEC >1 dictate further assessment on the effects of the compound on the fauna and flora within the environmental compartments. Studies to evaluate PEC values and, in turn, PEC/PNEC ratios for FQs are presently inadequate. However, an acceptable risk (PEC/PNEC <1) was reported for ofloxacin in water systems (Isidori et al. 2005). A study in Switzerland also suggested a low probability for adverse effects of ciprofloxacin and norfloxacin, either on microbial activity in WWTPs or on algae, *Daphnia*, and fish in surface water [risk quotient: MEC/PNEC <1 (Alder et al. 2003)]. In this review, an attempt was made to estimate and demonstrate the PEC value for difloxacin, taking the broiler chicken and laying hen as reference animals (Table 3). The PEC calculation is based on a worst case scenario assuming 100% excretion of the parent compound. The calculation is made with the help of a balancing model (Spaepen et al. 1997). More work in this field is very much needed.

XI. Conclusions

Scientific concern over the presence of antibiotics in different compartments of the environment is continuously increasing. The incidence of antibiotic resistance among strains of aquatic sources has already been reported (Schwartz et al. 2003; Messi et al. 2005). The presence of antibiotics in the aquatic environment and in soil systems has created concern about the potential toxicity of these compounds to nontarget aquatic organisms, soil microorganisms, and humans through drinking water. There is accumulating evidence of adverse human health consequences due to resistant organisms resulting from nonhuman usage of antimicrobials. Molecular characterization of resistance genes, with other data, indicates some movement of resistant bacteria and resistance determinants from aquaculture, companion animals, and horticulture to humans. The food-borne route is the major transmission pathway for resistant bacteria and resistance genes from food animals to humans, but other routes of transmission also exist.

It is necessary to determine the threshold concentration of antibiotics in soil, which may cause an increase in the percentage of resistant strains. Molecular techniques might give new insights into the provenance of resistant genes in soils and the possibility for horizontal gene transfer from resistant microorganisms to animal and human pathogens. There is a present-day need for realistic approaches in designing scientific studies on the behavior of FQs in the environment. Soil studies with FQs will remain incomplete when carried out only with pure active ingredient in the absence of manure. Again, spiking of the active drug to the manure does not provide answers on the metabolites generated in an animal system. The realistic approach should be to treat animals with FQs and study their fate and behavior in manure as such and also in soil, once mixed with that manure.

Table 3. Predicted Environmental Concentration (PEC) Calculation in Soil for Difloxacin when Applied to Broiler Chicken and Laying Hen.

	Calculated or suggested values		
Parameters	Broiler chicken	Laying hen	References
Weight of the animal (**BW**, kg)	1.3	2	Ref. 1
Animals raised per year on each place (**N**)	9	1	
ID (individual dose rate, mg/kg body wt.)*	30	30	
T (no. of individual treatments per animal)*	5	5	
Total amount of active ingredient (**Q**, mg/yr/place, $Q = ID \times BW \times T \times N$)	1,755	300	Ref. 1
Yearly output of excreta ($\mathbf{P_E}$, kg/place/yr)	37.2	67.5	Ref. 1
Concentration of active ingredient in liquid manure, mg/kg liquid manure ($\mathbf{C_E}$, $C_E = Q / P_E$)	47.18	4.44	Ref. 1
Maximum allowed nitrogen concentration ($\mathbf{A_N}$, kg/ha/yr)	170	170	Ref. 2
Annual nitrogen production per animal ($\mathbf{P_N}$, kg /place/yr)	0.32	0.32	Ref. 3
Maximum application of liquid manure/ha/yr (**M**, $M = A_N \times P_E / P_N$)	19762.5	35859.37	Ref. 1
Applied active ingredient, mg/ha/yr ($\mathbf{C_{SA}}$, $C_{SA} = M \times C_E$)	932,394.75	159,215.60	Ref. 1
W, kg/ha (Soil weight/ha/depth of input)**	3,750,000	3,750,000	
PEC in soil, mg/kg [$\mathbf{PEC = C_{SA} / (W + M)}$]	**0.247**	**0.042**	Ref. 1

*ID and T are considered as per the recommendation of Dicural, a commercially available formulation.

**calculation is based on soil depth of 25 cm and density of 1.5 g/cc.

Ref. 1, Spaepen et al. 1997; Ref. 2, Ministerium für Umwelt, und Naturschutz, Landwirtshaft und Verbraucherschutz des Landes Nordrhein-Westfalen, Düngeverordnung 2006; Ref. 3, Bilanzierung der Nährstoffausscheidungen Landwirtschatlicher Nutztiere,2005, Arbeiten der DLG, Band 199.

Unlike pesticides, the main route of entry of these compounds into the soil environment is through manure. The additional carbon source, by incorporating manure or sludge in soil, may influence the soil microbial biomass and, thus, promote or inhibit the dissipation of FQs in soil. Therefore, scientific approaches should be more rational and practical. From the present survey of the available literature it is clear that the fate and behavior of FQs and their metabolites in the environment is poorly understood. Therefore, it is indeed important to gain a better understanding on their persistence and geochemical transport for their environmental risk assessment. For this, studies with radioactive manure (manure obtained after feeding the animal with radioactive FQs) should be conducted to obtain data on the fate of antibiotics in manure and soil under realistic application and environmental conditions. Additionally, the normal practice of using OECD guideline no. 106 (OECD 2000) to determine the sorption coefficients of xenobiotics in soil does not reflect the influence of manure or sludge on the sorption of antibiotics. Thus, it is more rational to study the sorption behavior of FQs in soil in the presence of manure or sludge.

Research publications related to the presence of FQs in landfill leachates are scarce. This area should be considered in assessing overall environmental risks from FQs. Being antimicrobial, FQs may cause an alteration of the structural and functional diversity in the soil microbial population and, as a result, soil organic matter turnover and nutrient cycling is likely to be affected. Therefore, attempts should be made to minimize the loading of FQs in soil.

Summary

Fluoroquinolones (FQs) are used in large amounts for human and animal medical care. They are excreted as parent compound, as conjugates, or as oxidation, hydroxylation, dealkylation, or decarboxylation products of the parent compound. A considerable amount of FQs and their metabolites may reach the soil as constituents of urine, feces, or manure. The residues of FQs in foods of animal origin may pose hazards to consumers through emergence of drug-resistant bacteria. FQs bind strongly to topsoil, reducing the threat of surface water and groundwater contamination. The strong binding of FQs to soil and sediments delays their biodegradation and explains the recalcitrance of FQs. Wastewater treatment is an efficient elimination step (79%–87% removal) for FQs before they enter rivers. FQs are susceptible to photodegradation in aqueous medium, involving oxidation, dealkylation, and cleavage of the piperazine ring.

Acknowledgments

This study has been performed within the framework of the research project "Veterinary Medicines in Soils: Basic Research for Risk Analysis," funded by the German Science Foundation (Deutsche Forschungsgemeinschaft (DFG)), to whom the authors are deeply grateful. The authors thank Dr. Christian Corsing, Bayer AG, Monheim for his technical suggestions.

References

Ahel M, Jeličic I (2001) Phenazone analgesics in soil and ground water below a municipal solid waste landfill. In: Daughton CG, Jones-Lepp TL (eds) Pharmaceuticals and Personal Care Products in the Environment: Scientific and Regulatory Issues. Symposium Series 791. American Chemical Society, Washington, DC, pp 100–115.

Al-Ahmad A, Daschner FD, Kümmerer K (1999) Biodegradability of cefotiam, ciprofloxacin, meropenem, penicillin G and sulfamethoxazole and inhibition of waste water bacteria. Arch Environ Contam Toxicol 37:158–163.

Alder AC, McArdell CS, Golet EM, Ibric S, Molnar E, Nipales NS, Giger W (2001) Occurrence and fate of fluoroquinolone, macrolide, and sulfonamide antibiotics during wastewater treatment and in ambient waters in Switzerland. In: Daughton CG, Jones-Lepp TM (eds) Pharmaceuticals and Personal Care Products in the Environment: Scientific and Regulatory Issues. Symposium Series 791. American Chemical Society, Washington, DC, pp 56–69.

Alder AC, Golet EM, Giger W, Siegrist H (2003) Environmental exposure of fluoroquinolone antibacterial agents in wastewater and river water in Switzerland. Presented at ENVIRPHARMA, European Conference on Human and Veterinary Pharmaceuticals in the Environment, Lyon, France, April 14–16, 2003.

Alder AC, McArdell CS, Golet EM, Kohler HPE, Molnar E, Pham Thi NA, Siegrist H, Suter MJF, Giger W (2004) Environmental exposure of antibiotics in wastewaters, sewage sludges and surface waters in Switzerland. In: Kümmerer K (ed) Pharmaceuticals in the Environment: Sources, Fate, Effects and Risks, 2nd Ed. Springer-Verlag, Berlin, pp 55–66.

Andreozzi R, Rafaele M, Nicklas P (2003) Pharmaceuticals in STP effluents and their solar photodegradation in aquatic environment. Chemosphere 50:1319–1330.

Araki T, Kawai Y, Ohta I, Kitaoka H (2002) Photochemical behavior of sitafloxacin, fluoroquinolone antibiotic, in an aqueous solution. Chem Pharmaceut Bull 50: 229–234.

Asami S, Hatsugai K, Osawa K, Nakabayashi Y, Nakajima T (2000) Confirmation of quinolone antibacterials in porcine meat and kidneys by gas chromatography/mass spectrometry using selected ion monitoring. Nippon Juishikai Zasshi 53: 225–231.

Azerad R (1999) Microbial models for drug metabolism. Adv Biochem Eng Biotechnol 63:169–218.

Barbosa J, Barrón D, Jiménez-Lozano E, Sanz-Nebot V (2001) Comparison between capillary electrophoresis, liquid chromatography, potentiometric and spectrophotometric techniques for evaluation of pKa values of zwitterionic drugs in acetonitrile-water mixtures. Anal Chim Acta 437:309–321.

Barrón D, Jiménez-Lozano E, Bailac S, Barbosa J (2002) Determination of difloxacin and sarafloxacin in chicken muscle using solid-phase extraction and capillary electrophoresis. J Chromatogr B 767:313–319.

Barrón D, Jiménez-Lozano E, Bailac S, Barbosa J (2003) Simultaneous determination of flumequine and oxolinic acid in chicken tissues by solid-phase extraction and capillary electrophoresis. Anal Chim Acta 477:21–27.

Bhanot SK, Singh M, Chatterjee NR (2001) The chemical and biological aspects of fluoroquinolones: reality and dreams. Curr Pharm Design 7:313–337.

Bilanzierung der Nährstoffausscheidungen Landwirtschatlicher Nutztiere (2005) Arbeiten der DLG, Band 199.

Björklund H, Råbergh CMI, Bylund G (1991) Residues of oxolinic acid and oxytetracycline in fish and sediments from fish farms. Thesis. Department of Biology, Åbo Akademi University, Finland.

Booth DM (1994) Enrofloxacin revisited. Vet Med 8:744–753.

Borner K, Borner E, Hartwig H, Lode H (1990) HPLC of recent quinolone antimicrobials. In: Reid E, Wilson ID (eds) Analysis for Drugs and Metabolites, Including Antiinfective Agents. Royal Society of Chemistry, Cambridge, England, pp 131–144.

Boxall ABA, Fogg LA, Kay P, Blackwell PA, Pemberton EJ (2002) Review of veterinary medicines in the environment. Environment Agency R + D Technical Report P. Environment Agency, Bristol, UK.

Boxall ABA, Kolpin DW, Halling-Sørensen B, Tolls J (2003) Are veterinary medicines causing environmental risks? Environ Sci Technol 37:286–294.

Boxall ABA, Fogg LA, Blackwell PA, Kay P, Pemberton EJ, Croxford A (2004) Veterinary medicines in the environment. Rev Environ Contam Toxicol 180:1–91.

Bryskier A, Chantot JF (1995) Classification and structure–activity relationships of fluoroquinolones. Drugs 49:16–28.

Burhenne J, Ludwig M, Nikoloudis P, Spiteller M (1997a) Photolytic degradation of fluoroquinolone carboxylic acids in aqueous solution. Part I: primary photoproducts and half-lives. Environ Sci Pollut Res 4:10–15.

Burhenne J, Ludwig M, Spiteller M (1997b) Photolytic degradation of fluoroquinolone carboxylic acids in aqueous solution. Part II: Isolation and structural elucidation of polar photometabolites. Environ Sci Pollut Res 4:61–67.

Chen Y, Rosazza JPN, Reese CP, Chang HY, Nowakowski MA, Kiplinger JP (1997) Microbial models of soil metabolism: biotransformation of danofloxacin. J Ind Microbiol Biotechnol 19:378–384.

Chiu CH, Wu TL, Su LH (2002) The emergence in Taiwan of fluoroquinolone resistance in *Salmonella enterica serotype Choleraesuis*. N Engl J Med 346:413–419.

Choma IM (2003) TLC separation of fluoroquinolones: searching for better selectivity. J Liquid Chromatogr Relat Technol 26:2673–2685.

Choma IM, Choma A, Staszczuk K (2002) Determination of flumequine in milk by thin-layer chromatography-bioautography. J Liq Chromatogr Relat Technol 25:1579–1587.

Christian T, Schneider RJ, Färber HA, Skutlarek D, Meyer MT, Goldbach HE (2003) Determination of antibiotic residues in manure, soil, and surface waters. Acta Hydrochim Hydrobiol 31:36–44.

Cinquina AL, Roberti P, Giannetti L, Longo F, Draisci R, Fagiolo A, Brizioli NR (2003) Determination of enrofloxacin and its metabolite ciprofloxacin in goat milk by HPLC with diode-array detection. Optimization and validation. J Chromatogr A 987:221–226.

Dalhoff A, Bergan T (1998) Pharmacokinetics of fluoroquinolones in experimental animals. In: Kuhlmann J, Dalhoff A, Zeiler H-J (eds) Quinolone Antibacterials. Springer-Verlag, Berlin, pp 179–206.
Daughton CG, Ternes TA (1999) Pharmaceuticals and personal care products in the environment: agents of subtle change? Special report. Environ Health Perspect 107:907–938.
Davis IM, Lofthouse TJ, Stamm JM (1993) Aquatic photodegradation of ^{14}C-sarafloxacin hydrochloride. Abstr Pap Am Chem Soc 205:91.
Delepine B, Hurtaud-Pessel D, Sanders P (1998) Simultaneous determination of six quinolones in pig muscles by liquid chromatography-atmospheric pressure chemical ionisation mass spectrometry. Analyst 123:2743–2747.
Du LM, Lin AP, Yang YQ (2004a) Spectrofluorometric determination of certain quinolone through charge transfer complex formation. Anal Lett 37:2175–2188.
Du LM, Yang YQ, Wang QM (2004b) Spectrofluorometric determination of certain quinolone through charge transfer complex formation. Anal Chim Acta 516:237–243.
Edlund C, Lindqvist L, Nord CE (1988) Norfloxacin binds to human fecal material. Antimicrob Agents Chemother 32:1869–1874.
Ervik A, Thorsen B, Eriksen V, Lunestad BT, Samuelsen OB (1994) Impact of administering antibacterial agents on wild fish and blue mussels *Mytilus edulis* in the vicinity of fish farms. Dis Aquat Org 18:45–51.
Fasani E, Rampi M, Albini A (1999) Photochemistry of some fluoroquinolones: effect of pH and chloride ion. J Chem Soc Perkin Trans 2 9:1901–1907.
Fasani E, Mella M, Monti S, Albini A (2001) Unexpected photoreactions of some 7-amino-6-fluoroquinolones in phosphate buffer. Eur J Org Chem 2:391–397.
FEDESA (2001) Antibiotic use in farm animals does not threaten human health. FEDESA/FEFANA press release, European Federation of Animal Health, Brüssels, July 13, 2001.
FAO/OIE/WHO (2003) Evidence of association between non-human usage of antimicrobial agents and antimicrobial resistance in humans. In: Proceedings–Joint FAO/OIE/WHO Expert Workshop on Non-Human Antimicrobial Usage and Antimicrobial Resistance: Scientific assessment, December 1–5, 2003, Geneva, Switzerland, pp 11–15.
Garcia MA, Solans C, Calvo A, Hernandez E, Rey R, Bregante MA, Puig M (2005) Determination of enrofloxacin and its primary metabolite, ciprofloxacin, in pig tissues: application to residue studies. Biomed Chromatogr 19:27–31.
Golet EM, Alder AC, Hartmann A, Ternes TA, Giger W (2001) Trace determination of fluoroquinolone antibacterial agents in urban wastewater by solid-phase extraction and liquid chromatography with fluorescence detection. Anal Chem 73:3632–3638.
Golet EM, Alder AC, Giger W (2002a) Environmental exposure and risk assessment of fluoroquinolone antibacterial agents in wastewater and river water of the Glatt Valley Watershed, Switzerland. Environ Sci Technol 36:3645–3651.
Golet EM, Strehler A, Alder AC, Giger W (2002b) Determination of fluoroquinolone antibacterial agents in sewage sludge and sludge-treated soil using accelerated solvent extraction followed by solid-phase extraction. Anal Chem 74: 5455–5462.

Golet EM, Xifra I, Siegrist H, Alder AC, Giger W (2003) Environmental exposure assessment of fluoroquinolone antibacterial agents from sewage to soil. Environ Sci Technol 37:3243–3249.

Grellet J, Ba B, Saux MC (2002) High-performance liquid chromatographic separation of fluoroquinolone enantiomers: a review. J Biochem Biophys Methods 54:221–233.

Griffin DM (1972) Ecology of Soil Fungi. Chapman & Hall, London.

Halling-Sørensen B, Nielsen SN, Lanzky PF, Ingerslev F, Holten Lützhoft HC, Jorgensen SE (1998) Occurrence, fate and effect of pharmaceutical substances in the environment: a review. Chemosphere 36:357–393.

Halling-Sørensen B, Holten Lützhoft HC, Andersen HR, Ingerslev F (2000) Environmental risk assessment of antibiotics: comparison of mecillinam, trimethoprim and ciprofloxacin. J Antimicrob Chemother 46:53–58.

Halling-Sørensen B, Sengeløv G, Ingerslev F, Jensen LB (2003) Reduced antimicrobial potencies of oxytetracycline, tylosin, sulfadiazine, streptomycin, ciprofloxacin, and olaquindox due to environmental processes. Arch Environ Contam Toxicol 44:7–16.

Hamscher G, Sczesny S, Abu-Quare A, Hörper H, Nau H (2000) Substances with pharmacological effects including hormonally active substances in the environment: identification of teteracyclines in soil fertilized with animal slurry. Dtsch Tierärztl Wochenschr 107:293–348.

Hamscher G, Powelzick HT, Höper H, Nau H (2004) Antibiotics in soil: routes of entry, environmental concentrations, fate and possible effects. In: Kümmerer K (ed) Pharmaceuticals in the Environment: Sources, Fate, Effects and Risks, 2nd Ed. Springer-Verlag, Berlin, pp 139–147.

Hartmann A, Alder AC, Koller T, Widmer RM (1998) Identification of fluoroquinolone antibiotics as the main source of *umuC* genotoxicity in native hospital wastewater. Environ Toxicol Chem 17:377–382.

Hartmann A, Golet EM, Gartiser S, Alder AC, Koller T, Widmer RM (1999) Primary DNA damage but not mutagenecity correlates with ciprofloxacin concentrations in German hospital wastewaters. Arch Environ Contam Toxicol 36:115–119.

Heberer T (2002) Tracking persistent pharmaceutical residues from municipal sewage to drinking water. J Hydrol 266:175–189.

Hektoen H, Berge JA, Hormazabal V, Yndestad M (1995) Persistence of antibacterial agents in marine sediments. Aquaculture 133:175–184.

Hernández M, Borrull F, Calull M (2002) Using non-aqueous capillary electrophoresis to analyze several quinolones in pig kidney samples. Electrophoresis 23:506–511.

Holm JV, Ruegge K, Bjerg PL, Christensen TH (1995) Occurrence and distribution of pharmaceutical organic compounds in the groundwater downgradient of a landfill (Grinsted, Denmark). Environ Sci Technol 29:1415–1420.

Hooper DC, Wolfson JS (1993) Mechanism of quinolone action and bacterial killing. In: Hooper DC, Wolfson JS (eds) Quinolone Antibacterial Agents, 2nd Ed. American Society for Microbiology, Washington DC, pp 53–75.

Ingerslev F, Torang L, Loke ML, Halling-Sørensen B, Nyholm N (2001) Primary biodegradation of veterinary antibiotics in aerobic and anaerobic surface water simulation systems. Chemosphere 44:865–872.

Isidori M, Lavorgna M, Nardelli A, Pascarella L, Parrella A (2005) Toxic and genotoxic evaluation of six antibiotics on non-target organisms. Sci Total Environ 346:87–98.

Kassenborg HD, Smith KE, Vugia DJ, Rabatsky-Ehr T, Bates MR, Carter MA, Dumas NB, Cassidy MP, Marano N, Tauxe RV, Angulo FJ (2004) Fluoroquinolone-resistant *Campylobacter* infections: eating poultry outside of the home and foreign travel are risk factors. Clin Infect Dis 38:279–284.
Kay P, Blackwell PA, Boxall ABA (2004) Fate of veterinary antibiotics in a macroporous tile drained clay soil. Environ Toxicol Chem 23:1136–1144.
Kolpin DW, Furlong ET, Meyer MT, Thurman EM, Zaugg SD, Barber LB, Buxton HT (2002) Pharmaceuticals, hormones, and other organic wastewater contaminants in U. S. streams, 1999–2000: a national reconnaissance. Environ Sci Technol 36:1202–1211.
Kümmerer K (2004) Emissions from medical care units. In: Kümmerer K (ed) Pharmaceuticals in the Environment: Sources, Fate, Effects and Risks, 2nd Ed. Springer-Verlag, Berlin, pp 27–44.
Kümmerer K, Al-Ahmed A, Mersch-Sundermann V (2000) Biodegradability of some antibiotics, elimination of their genotoxicity and affection of waste water bacteria in a simple test. Chemosphere 40:701–710.
Lode H, Höffken G, Prinzing C, Glatzel P, Wiley R, Olschewski P, Sievers B, Reimnitz D, Borner K, Koeppe P (1987) Comparative pharmacokinetics of new quinolones. Drugs 34:21–25.
Lunestad BT, Samuelsen OB, Fjelde S, Ervik A (1995) Photostability of eight antibacterial agents in seawater. Aquaculture 134:217–225.
Lützhøft H-CH, Vaes WHJ, Freidig AP, Halling-Sørensen B, Hermens JLM (2000) Influence of pH and other modifying factors on the distribution behavior of 4-quinolones to solid phases and humic acids studied by "negligible-depletion" SPME-HPLC. Environ Sci Technol 34:4989–4994.
Marazuela MD, Moreno-Bondi MC (2004) Multiresidue determination of fluoroquinolones in milk by column liquid chromatography with fluorescence and ultraviolet absorbance detection. J Chromatogr A 1034:25–32.
Marengo JR, Kok RA, O'Brian K, Velagaleti RR, Stamm JM (1997) Aerobic biodegradation of (^{14}C)-sarafloxacin hydrochloride in soil. Environ Toxicol Chem 16:462–471.
Marengo JR, Kok RA, Burrows LA, Velagaleti RR, Stamm JM (2001) Biodegradation of (^{14}C)-sarafloxacin hydrochloride, a fluoroquinolone antimicrobial, by *Phanerochaete chrysosporium*. J Sci Ind Res 60:121–130.
Matzen EL, Wammer KH, Werner JH, McNeil K (2005) Aquatic environmental photochemistry and biological activity of fluoroquinolone antibiotics. In: Abstracts of Papers, 229th ACS National Meeting, San Diego, CA, March 13–17, 2005.
Mella M, Fasani E, Albini A (2001) Photochemistry of 1-cyclopropyl-6-fluoro-1, 4-dihydro-4-oxo-7-(piperazin-1-yl) quinoline-3-carboxylic acid (ciprofloxacin) in aqueous solutions. Helv Chim Acta 84:2508–2519.
Messi P, Guerrieri E, Bondi M (2005) Antibiotic resistance and antibacterial activity in heterotrophic bacteria of mineral water origin. Sci Total Environ 346:213–219.
Miao X-S, Bishay F, Chen M, Metcalfe CD (2004) Occurrence of antimicrobials in the final effluents of wastewater treatment plants in Canada. Environ Sci Technol 38:3533–3541.
Ministerium für Umwelt, and Naturschutz, Landwirtshaft und Verbraucherschutz des Landes Nordrhein-Westfalen (2006) Düngeverordnung.

Molbak K, Baggesen DL, Aarestrup FM, Ebbesen JM, Engberg J, Frydendahl K, Gerner-Smidt P, Petersen AM, Wegener HC (1999) An out-break of multi-drug resistant, quinolone-resistant *Salmonella enterica serotype Typhimurium* DT 104. N Engl J Med 341:1420–1425.

Morais-Cabral JH, Jackson AP, Smith CV, Shikotra N, Maxwell A, Liddington RC (1997) Crystal structure of breakage-reunion domain of DNA gyrase. Nature (Lond) 388:903–906.

Nakamura R, Yamaguchi T, Sekine Y, Hashimoto M (1983) Determination of new antibacterial agent (AT-2266) and its metabolites in plasma and urine by high-performance liquid chromatography. J Chromatogr 278:321–328.

Nakashima M, Kanamaru M, Uematsu T, Takiguchi A, Mizuno A, Itaya T, Kawahara F, Ooie T, Saito S, Uchida H, Masuzawa K (1988) Clinical pharmacokinetics and tolerance of fleroxacin in healthy male volunteers. J Antimicrob Chemother 22:133–144.

Nakata H, Kannan K, Jones PD, Giesy JP (2005) Determination of fluoroquinolone antibiotics in wastewater effluents by liquid chromatography–mass spectrometry and fluorescence detection. Chemosphere 58:759–766.

Nowara A, Burhenne J, Spiteller M (1997) Binding of fluoroquinolone carboxylic acid derivatives to clay. J Agric Food Chem 45:1459–1463.

OECD (2000) Adsorption/desorption using a batch equilibrium method. OECD guidelines for testing of chemicals: test guideline 106. Revised draft document. OECD, Paris, France, pp 1–26.

Ohlsen K, Ziebuhr W, Koller K, Hell W, Wichelhaus TA, Hacker J (1998) Effects of sub-inhibitory concentrations of antibiotics an alpha-toxon (hla) gene expression of methicillin-sensitive and methicillin-resistant *Staphylococcus aureus* isolates. Antimicrob Agents Chemother 42:2817–2823.

Parshikov IA, Freeman JP, Lay JO, Beger RD, Williams AJ, Sutherland JB (1999) Regioselective transformation of ciprofloxacin to N-acetylciprofloxacin by the fungus *Mucor ramannianus.* FEMS Microbiol Lett 177:131–135.

Parshikov IA, Freeman JP, Lay JO, Beger RD, Williams AJ, Sutherland JB (2000) Microbiological transformation of enrofloxacin by the fungus *Mucor ramannianus*. Appl Environ Microbiol 66:2664–2667.

Pecorelli I, Galarini R, Bibi R, Floridi Al, Casciarri E, Floridi A (2003) Simultaneous determination of 13 quinolones from feeds using accelerated solvent extraction and liquid chromatography. Anal Chim Acta 483:81–89.

Pfenning AP, Munnas RK, Turnipseed SB, Roybal JE, Holland DC, Long AR, Plakas SM (1996) Determination and confirmation of identities of flumequine, nalidixic, oxolinic and piromidic acids in salmon and shrimp. J AOAC Int 79:1227–1235.

Pham Thi NA (2003) Determination and occurrence of fluoroquinolone antibacterial agents in hospital waste water. EPFL, Lausanne, Switzerland.

Philips G, Jhonson BE, Ferguson J (1990) The loss of antimicrobial activity of ciprofloxacin by photodegradation. J Antimicrob Chemother 26:783–789.

Renew JE, Huang CH (2004) Simultaneous analysis of fluoroquinolone, sulfonamide and trimethoprim antibiotics in wastewater using tandem solid-phase extraction and liquid chromatography electrospray mass spectrometry. J Chromatogr A 1042:113–121.

Samanidou VF, Christodoulou EA, Papadoyannis IN (2005a) Determination of fluoroquinolones in edible animal tissue samples by high performance liquid chromatography after solid phase extraction. J Sep Sci 28:555–565.

Samanidou VF, Christodoulou EA, Papadoyannis IN (2005b) Direct determination of five fluoroquinolones in chicken whole blood and in veterinary drugs by HPLC. J Sep Sci 28:325–331.
Schneider MJ, Donoghue DJ (2003) Multiresidue determination of fluoroquinolone antibiotics in eggs using liquid chromatography-fluorescence-mass spectrometry. Anal Chim Acta 483:39–49.
Schwartz T, Kohnenb W, Jansenb B, Obsta U (2003) Detection of antibiotic-resistant bacteria and their resistance genes in wastewater, surface water, and drinking water biofilms. FEMS Microbiol Ecol 43:325–335.
Shen JY, Kim MR, Lee CJ, Kim IS, Lee KB, Shim JH (2004) Supercritical fluid extraction of the fluoroquinolones, norfloxacin and ofloxacin from orally treated chicken breast muscles. Anal Chim Acta 513:451–455.
Shim JH, Shen JY, Kim MR, Lee CJ, Kim IS (2003) Determination of the fluoroquinolone enrofloxacin in edible chicken muscle by supercritical fluid extraction and liquid chromatography with fluorescence detection. J Agric Food Chem 51:7528–7532.
Sithole BB, Guy RD (1987) Models for tetracycline in aquatic environments. I. Interaction with bentonite clay systems. Water Air Soil Pollut 32:303–314.
Snyder SA, Westerhoff P, Yoon Y, Sedlak DL (2003) Pharmaceuticals, personal care products and endocrine disrupters in water: implications for water treatment. Environ Eng Sci 20:449–469.
Sörgel F, Kinzig M (1993) Pharmacokinetics of gyrase inhibitors. Part 2: Renal and hepatic elimination pathways and drug interactions. Am J Med 94:56–69.
Sørum H, L'Abée-Lund TM (2002) Antibiotic resistance in food-related bacteria a result of interfering with the global web of bacterial genetics. Int J Food Microbiol 78:43–56.
Spaepen KRI, Van Leemput LJJ, Wislocki PG, Verschueren C (1997) A uniform procedure to estimate the predicted environmental concentration of the residues of veterinary medicines in soil. Environ Toxicol Chem 16:1977–1982.
Stamatelatou K, Lyberatos G, Drillia P (2003) The fate of pharmaceuticals on soil. Presented at ENVIRPHARMA, European Conference on Human and Veterinary Pharmaceuticals in the Environment, Lyon, France, April 14–16, 2003.
Sudo K, Okazaki O, Tsumura M, Tachizawa H (1986) Isolation and identification of metabolites of ofloxacin in rats, dogs, and monkeys. Xenobiotica 16:725–732.
Sunderland J, Tobin CM, White LO, MacGowan AP, Hedges AJ (1999) Ofloxacin photodegradation products possess antimicrobial activity. Drugs 58:171–172.
Sunderland J, Tobin CM, Hedges AJ, MacGowan AP, White LO (2001) Antimicrobial activity of fluoroquinolone photodegradation products determined by parallel-line bioassay and high performance liquid chromatography. J Antimicrob Chemother 47:271–275.
Thiele-Bruhn S (2003) Pharmaceutical antibiotic compounds in soils—a review. J Plant Nutr Soil Sci 166:145–167
Thurman EM, Dietze LE, Scribner EA (2002) Occurrence of antibiotics in water from fish hatcheries. Fact Sheet 120–02. U.S. Geological Survey, Washington, DC.
Tolls J (2001) Sorption of veterinary pharmaceuticals in soils: a review. Environ Sci Technol 35:3397–3406.
Tyczkowska K, Hedeen KM, Aucoin DP, Aronson AL (1989) High-performance liquid chromatographic method for the simultaneous determination of enro-

floxacin and its primary metabolite ciprofloxacin in canine serum and prostatic tissue. J Chromatogr 493:337–346.
Van Dijk J, Keukens HJ (2000) The stability of some veterinary drugs and coccidiostats during composting and storage of laying hen and broiler feces. In: Van Ginkel LA, Ruiter A (eds) Residues of Veterinary Drugs in Food. Proceedings of the Euroresidue IV Conference, Veldhoven, The Netherlands, May 8–10, 2000, pp 356–360.
Van Hoof N, Wasch KD, Okerman L, Reybroeck W, Poelmans S, Noppe H, Brabander HD (2005) Validation of a liquid chromatography-tandem mass spectrometric method for the quantification of eight quinolones in bovine muscle, milk and aquacultured products. Anal Chim Acta 529:265–272.
Van Saene JJM, Van Saene HKF, Lerk CF (1986) Inactivation of quinolone by feces. J Infect Dis 153:999–1000.
Velagaleti RR, Davis ML, O'Brien GK, Stamm JM (1993) The bioavailability of ^{14}C-sarafloxacin hydrochloride in three soils and a marine sediment as determined by biodegradation and sorption/desorption parameters. In: Abstracts of the 205th National Meeting of the American Chemical Society, Division of Agrochemicals, American Chemical Society, Washington, DC, Abstract 92.
Vílchez JL, Araujo L, Prieto A, Navalón A (2004) Determination of ciprofloxacin and enoxacin in human serum samples by micellar liquid chromatography. Anal Chim Acta 516:135–140.
Walker RA, Lawson AJ, Lindsay EA, Ward LR, Wright PA, Bolton FJ, Wareing DR, Corkish JD, Davies RH, Threlfall EJ (2000) Decreased susceptibility to ciprofloxacin in outbreak-associated multi-resistant *Salmonella typhimurium* DT 104. Vet Rec 147:395–396.
Wang PL, Chen L, Fan YF (2001) Selective separation and simultaneous determination of trace levels of five types of fluorinated quinolone drugs by thin-layer chromatography/fluorescence densiometry. J AOAC Int 84:684–688.
Wetzstein HG (2001) Chinolone in der Umwelt: Biologische Abbaubarkeit der Gyrasehemmer. Pharm Unserer Zeit 30:450–457.
Wetzstein H-G, Schmeer N, Karl W (1997) Degradation of fluoroquinolone enrofloxacin by the brown rot fungus *Gloeophyllum striatum*: identification of metabolites. Appl Environ Microbiol 63:4272–4281.
Wetzstein H-G, Stadler M, Tichy H-V, Dalhoff A, Karl W (1999) Degradation of ciprofloxacin by basidomycetes and identification of metabolites generated by the brown rot fungus *Gloeophyllum striatum*. Appl Environ Microbiol 65:1556–1563.
Wetzstein H-G, Karl W, Hallenbach W, Himmler T, Petersen U (2000) Residual antibacterial activity of metabolites derived from the vetrinary fluoroquinolone enrofloxacin. 100th General Meeting of the American Society for Microbiology, Los Angeles, CA.
WHO (1998) Use of quinolones in food animals and potential impact on human health. WHO/EMC/ZDI/98.10. Report of a WHO Meeting, Geneva, Switzerland, June 2–5, 1998.
Williams AJ, Parshikov IA, Moody JD, Heinze TM, Sutherland JB (2004) Fungal transformation of an antimicrobial fluoroquinolone drug during growth on poultry litter materials. J Appl Poult Res 13:235–240.
Wise R (2002) Antimicrobial resistance: priorities for action. J Antimicrob Chemother 49:585–586.

Xifra I (2000) Behaviour of ciprofloxacin and norfloxacin during wastewater treatment. ETH, Zurich, Switzerland.

Zeiler H-J, Peterson U, Gau W, Ploschke HJ (1987) Antibacterial activity of the metabolites of ciprofloxacin and its significance in the bioassay. Arzneim Forschung/Drug Res 37:131–134.

Zhang H, Huang C-H (2005) Oxidative transformation of fluoroquinolone antibacterial agents and structurally related amines by manganese oxide. Environ Sci Technol 39:4474–4483.

Zuccato E, Castiglioni S, Fanelli R, Bagnati R, Calamari D (2004) Pharmaceuticals in the environment: changes in the presence and concentrations of pharmaceuticals for human use in Italy. In: Kümmerer K (ed) Pharmaceuticals in the Environment: Sources, Fate, Effects and Risks, 2nd Ed. Springer-Verlag, Berlin, pp 45–53.

Manuscript received September 6; accepted September 14, 2006.

Rev Environ Contam Toxicol 191:163–215

Explosives: Fate, Dynamics, and Ecological Impact in Terrestrial and Marine Environments

Albert L. Juhasz and Ravendra Naidu

Contents

I. Introduction

Explosive, or energetic compounds, may be defined as chemicals that, under the influence of thermal or chemical shock, decompose rapidly with the evolution of large amounts of heat and gas (Brannon and Pennington 2002). Numerous energetic compounds have been produced for varying industrial

---

Communicated by David M. Whitacre.

A. L. Juhasz (✉), R. Naidu
Centre for Environmental Risk Assessment and Remediation, University of South Australia, and The Cooperative Research Centre for Contamination Assessment and Remediation of the Environment (CRC CARE), Mawson Lakes Campus, Adelaide, Australia, 5095

uses; however, secondary explosives pose the largest potential environmental concern because they are produced and used in defense activities in the greatest quantities. Secondary explosives may enter the environment following explosives manufacture, assembly, and packing, and explosives detonation. During these activities, soil, sediment, and water may become contaminated with energetic and related compounds with potential impacts on environmental and human health. Of the secondary explosives, trinitrotoluene (TNT) and Royal Demolition Explosive (hexahydro-1,3,5-trinitro-1,3,5-triazine) (RDX) production outweigh other secondary explosives as they are the major ingredients in nearly every munition formulation (Walsh et al. 1993). In addition to chemicals added to explosive formulations, residues may contain compounds such as production impurities or decomposition by-products. For example, High Melting Explosive (octahydro-1,3,5,7-tetranitro-1,3,5,7-tetrazocine) (HMX) may be found as an impurity in RDX (Army, U.S. Department of Defense 1994), and TNT may contain dinitrotoluene and trinitrotoluene isomers (Legett et al. 1977).

A considerable amount of research has investigated the fate and dynamics of energetic compounds, focusing on contaminants in terrestrial and freshwater environments. Although less research has focused toward the ecological impact of energetic compounds, especially in the marine environment, identification and validation of test organisms for ecological risk assessment is an emerging research area. This review provides an outline of biotic and abiotic processes influencing the fate and dynamics of TNT, RDX, and HMX in the environment and the impact of these compounds on aquatic, marine, and terrestrial receptors.

II. Physicochemical Properties of Energetic Compounds

There are three main categories for organic secondary energetic materials: nitroaromatics, nitramines, and nitrate esters. Nitroaromatics, such as TNT, ammonium picrate, and methyl-2,4,6-trinitrophenylnitramine (Tetryl), contain NO_2 groups that are bonded to carbon atoms on an aromatic ring. Nitramines contain NO_2 groups that are bonded to a nitrogen atom present within an alicyclic ring, e.g., RDX and HMX; nitrate esters contain NO_2 groups bonded to an oxygen atom attached to an aliphatic carbon, e.g., pentaerythritol-tetranitrate (PETN).

Other energetic compounds may be classified as inorganic compounds (e.g., lead azide and ammonium nitrate), nitroso compounds (e.g., tetrazene), metallic derivatives (e.g., mercury fulminate and lead styphnate), or mixtures of oxidisable materials (e.g., fuels and oxidizing agents).

A. TNT

TNT (Fig. 1) is one of the most common bulk explosives used in military and civilian activities. It has been mixed with other energetic compounds

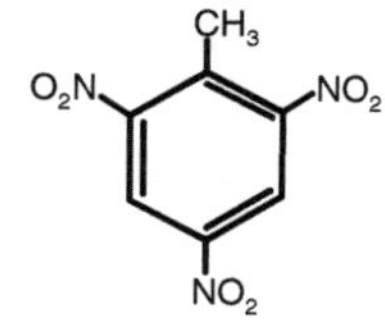

Trinitrotoluene (TNT)

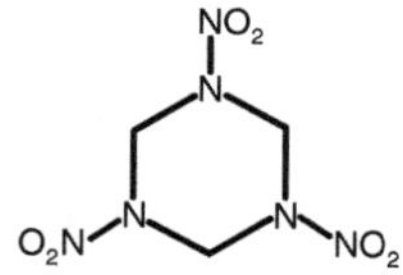

Royal Demolition Explosive
(RDX)

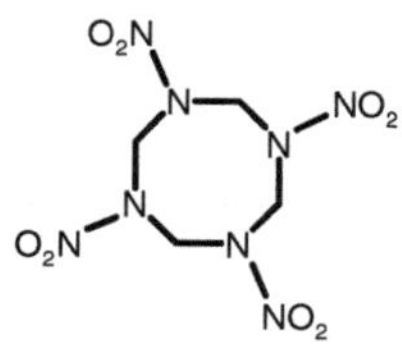

High Melting Explosive
(HMX)

Fig. 1. Chemical structures of trinitrotoluene (TNT), Royal Demolition Explosive (RDX), and High Melting Explosive (HMX).

(e.g., RDX and HMX) to form explosive composites (Table 1). TNT is chemically and thermally stable, has a low melting point, favorable for melt casting operations, and is of moderate toxicity. The nitroaromatic is sparingly soluble in water and has a low vapour pressure and Henry's law constant, indicating low volatility (Table 2). The octanol–water partitioning coefficient (K_{ow} of 1.86) of TNT indicates that, once dissolved, TNT will not sorb strongly to soil/sediment and therefore may be mobile in the environment.

Dinitrotoluene (DNT) isomers (2,4-DNT and 2,6-DNT) may occur as impurities during the manufacture of TNT or may be formed during the biotic and abiotic transformation of TNT. 2,4- and 2,6-DNT have similar chemical properties (see Table 2): they have a low aqueous solubility, are relatively nonvolatile, and have octanol–water partitioning coefficients of 1.98 and 2.02, respectively.

2-Amino-4,6-dinitrotoluene (2-A-4,6-DNT) and 4-amino-2,6-dinitrotoluene (4-A-2,6-DNT) may occur in the environment as a result of TNT transformation. These compounds are produced as a result of the biotic transformation of nitro functional groups to amino groups. Amino dinitrotoluene (2-A-4,6-DNT, 4-A-2,6-DNT) isomers have similar chemical properties (see Table 2): they are relatively nonvolatile and have a solubility of

Table 1. Composition of Common Secondary Explosives.

Name	Composition
Amatex	TNT, ammonium nitrate, RDX
Ammonal	TNT, ammonium nitrate, aluminium
Anatols	TNT, ammonium nitrate
Baratol	TNT, barium nitrate
C-4	RDX (91%), plasticizer (9%)
Composition A	RDX (91%), wax (9%)
Composition B	RDX (60%), TNT (39%), wax (1%)
Cycolotol	RDX, TNT
Explosive D	Ammonium picrate, picric acid
HTA-3	HMX, TNT, aluminium
Minol	TNT, ammonium nitrate, aluminium
Octol	HMX (70%–75%), TNT (25%–30%)
Pentolite	Ammonium picrate, TNT
Tetryltols	Tetryl, TNT
Torpex	RDX, TNT, aluminium
Tritonal	TNT (80%), aluminium (20%)

Source: Modified from Walsh et al. (1993).

17 and 36 mg/L, respectively. Although amino dinitrotoluenes have a low octanol–water partitioning coefficient (K_{ow} of 2.8 and 2.62), they may bind covalently to organic and mineral components in soil and sediment.

B. RDX

RDX (see Fig. 1) is a highly stable nitramine compound and is considered the most powerful and brisant of the military high explosives. RDX is sparingly soluble in water and has a low vapor pressure, indicating low volatility (see Table 2). The Henry's constant value of $>6.3 \times 10^{-8}$ atm-m^3 mol^{-1} indicates that RDX will not readily volatilize from aqueous solutions. The RDX octanol–water partitioning coefficient (K_{ow}, 0.86) indicates that, once dissolved, RDX will not sorb strongly to soil or sediment.

C. HMX

HMX (see Fig. 1) is used exclusively for military purposes as a component of plastic bonded explosives in rocket propellant and as a high explosive bursting charge. The nitramine is not volatile, has a water solubility of approximately 5 mg/L, and an octanol–water partitioning coefficient of 0.061 (see Table 2). The hydrophobicity of HMX does not limit its solubility, rather its high crystal energy. Once dissolved, HMX does not readily sorb to soil/sediment and therefore may be mobile in the environment.

Table 2. Physical and Chemical Properties of TNT, RDX, HMX, and TNT Transformation Products.

Energetic Compound	TNT	RDX	HMX	2,4-DNT	2,6-DNT	2-A-4,6-DNT	4-A-2,6-DNT
CAS number	38082-89-2	00121-82-4	026914-41-0	121-14-2	606-20-2	35572-78-2	19406-51-0
Synonyms and commercial names	2,4,6-Trinitrotoluene, trinitrotol, trilite, tolite, trinol, trotyl, tritolol, tritone, trotol, triton	Cyclotrimethylene-trinitramide, cyclonite, hexogen, composition A-6	Cyclotetromethylene-tetranitramine, octogen, octahydro-1,3,5,7-tetranitro-1,3,5,7-tetrazocine	2,4-Dinitrotoluene	2,6-Dinitrotoluene	2-Amino-4,6-dinitrotoluene	4-Amino-2,6-dinitrotoluene
Chemical formula	$C_7H_5N_3O_6$	$C_3H_6N_6O_6$	$C_4H_8N_8O_8$	$C_7H_6N_2O_4$	$C_7H_6N_2O_4$	$C_7H_7N_3O_4$	$C_7H_7N_3O_4$
Melting point	80°–82°C	204°C	276°–280°C	70°C	64°–66°C	176°C	171°C
Solubility in water	130 mg/L @ 20°C	42 mg/L @ 20°C	5 mg/L @ 25°C	270 mg/L @ 22°C	206 mg/L @ 25°C	17 mg/L	36 mg/L
Specific gravity (g/mL)	1.5–1.6	1.82	1.96	1.32	1.28	No data	No data
Vapor pressure (atm @ 20°C)	7.2×10^{-9}	5.3×10^{-12}	4.3×10^{-17}	2.9×10^{-7} @ 25°C	7.5×10^{-7}	5.3×10^{-8}	2.6×10^{-8}
Henry's law constant (atm-m³/mol)	4.57×10^{-7} to 1.1×10^{-8}	6.3×10^{-8} to 1.96×10^{-11}	2.6×10^{-15}	$1.3 - 1.86 \times 10^{-7}$	4.86×10^{-7} to 9.26×10^{-8}	1.19×10^{-7}	1.74×10^{-7}
Octanol–water partitioning coefficient	1.86	0.86	0.061	1.98	2.02	2.8	2.62
Appearance and odor	Yellow flakes with bitter almond odor	White or gray powder, odorless	White or gray powder, odorless	Yellow needles or monoclinic prisms	Yellow to red rhombic needles	—	—

Source: McGrath (1995); Brannon and Pennington (2002); Material Safety Data Sheets.

III. Energetics: Fate and Behavior

A. Source and Levels of Energetic Compounds in the Environment

Exposure assessment and risk management of energetics-contaminated soil, sediment, and water requires the knowledge of the fate and effect of energetics and their transformation products in the environment (Brannon and Myers 1997). TNT, RDX, and HMX may enter the environment via a number of sources:

- Production facilities, e.g., wastewater lagoons, filtration pits
- Solid waste destruction facilities, e.g., burn pits, incineration waste
- Packing or warehouse facilities
- Dispersed or unexploded ordnances, e.g., firing ranges

During their detonation, energetic compounds may undergo high-order or low-order detonation. Complete or high-order detonation causes a "kick out" of both munition debris and small quantities of munition constituents (energetic compounds and metals) into the environment. Incomplete or low-order detonation causes a "kick out" of not only munitions debris and large quantities of munitions constituents, but also larger pieces of the munition itself, into the environment. Low-order detonation often results in energetic compounds being present in soil and sediments as discrete particles. If the ordnance fails to fire (i.e., unexploded ordnance, UXO), the energetic compound may be released to the environment over time if corrosion of the shell occurs.

Most site assessments and remediation efforts have been directed toward the soil environment because spill sites, disposal areas, and military ranges are predominantly located in upland areas. In addition, groundwater contamination assessments have been performed because groundwater issues are associated with wastewater lagoons and leach pits located at production and packing facilities. Table 3 outlines the range of TNT, RDX, HMX, and TNT transformation product concentrations found during investigation of potentially contaminated military facilities, firing ranges, and disposal areas. Few studies have assessed the occurrence of energetic compounds in the marine environment. Of the studies undertaken in the marine environment (Ampleman et al. 2004; Carr and Nipper 2003; MLA 1996), the focus has been on the collection of water and sediment samples before and after detonation of UXOs. Although energetic compounds were not detected at these sites, the authors emphasized the difficulties associated with sampling and assessing the environmental fate of underwater UXOs.

The environmental fate and potential hazard of energetic compounds in the environment is affected by a number of processes including dissolution, sorption, volatiliation, abiotic and biotic transformation/degradation and bioaccumulation (Pennington and Brannon 2002) (Fig. 2). The following subsections describe fate and transport processes affecting energetic compounds in the environment.

Table 3. Concentration of Energetic Compounds in Contaminated Soils and Sediments.

Location / matrix	Energetic Compound (mg/kg)							Notes	Reference
	TNT	2,4-DNT	2,6-DNT	2A-DNT	4A-DNT	RDX	HMX		
Terrestrial environment:									
Fort Greely, Alaska	0.0058	—	—	0.002	0.003	0.94	0.22	Soil sampled underneath an 81-mm projectile low-order detonation.	Walsh et al. (2001)
Fort Greely, Alaska	130	0.036	0.016	0.84	1.0	340	40	Sample collected from underneath a 2.75-inch rocket low-order detonation.	Walsh et al. (2001)
Donnelly Training Area, Alaska	314	123	—	—	—	1.4	0.11	Soil collected from a target array.	Walsh et al. (2004)
Fort Ord, USA	0.17	—	0.9	1.08	—	0.5	587	Soil collected from a firing range.	Jenkins et al. (1998)
Umatilla, Oregon	1869	—	—	—	—	1069	175	—	Weston (1993)
Marine environment:									
Port Amour, Canada	n.d.	n.d.	n.d.	n.d.	n.d.	n.d.	n.d.	Analysis of water and sediment close to UXOs.	Ampleman et al. (2004)
Halifax Harbour, Canada	n.d.	n.d.	n.d.	n.d.	n.d.	n.d.	n.d.	Analysis of water and sediment close to UXOs. Pb was detected in sediments.	Ampleman et al. (2004)
Beaufort's Dyke disposal site, North Irish Sea	n.d.	n.d.	n.d.	n.d.	n.d.	n.d.	n.d.		MLA (1996)

UXO, unexploded ordnance.
Note the Range of Concentrations Detected in Soils and the Limited Number of Marine Sediment Assessments.

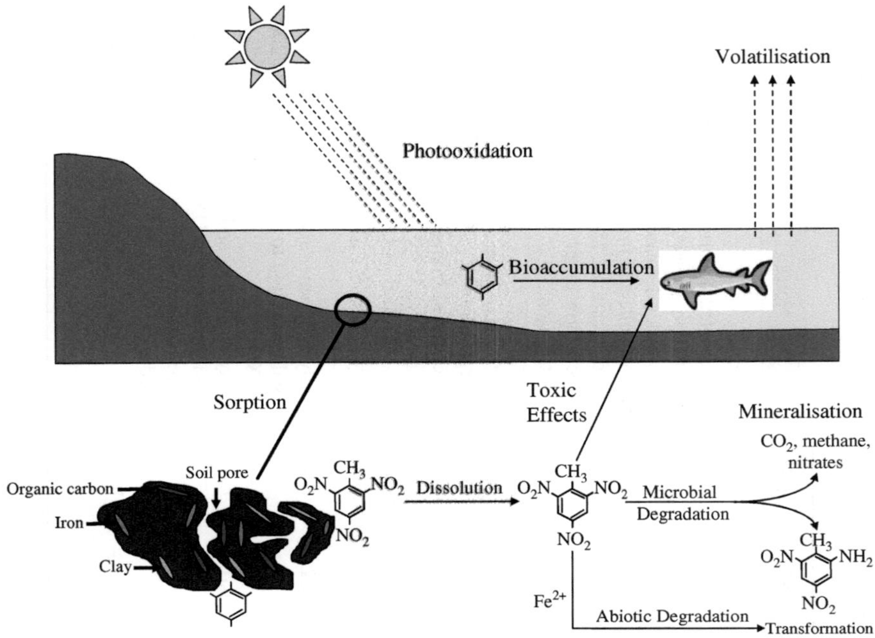

Fig. 2. Fate of energetic compounds in the environment.

B. Solubility and Dissolution

A critical parameter in determining the environmental fate and transport of energetic compounds is the aqueous solubility. Dissolution into water is a primary mechanism by which solid energetic compounds are disseminated throughout the environment (Lynch 2002). Once dissolved, energetic compounds may be transferred to other matrices (via adsorption or uptake) or undergo transformation by biotic or abiotic process. As seen in Table 2, TNT, RDX, and HMX are characterized by their low aqueous solubility (5–130 mg/L). If discrete energetic particles are present as a result of low-order detonations, the energetic compound may dissolve slowly over time.

Few studies have determined the rate of energetic dissolution. In dissolution studies performed by Lynch (2002) and Lynch et al. (2002) with TNT, HMX, and RDX in mixed aqueous systems, the rate of energetic dissolution followed the order TNT > HMX > RDX with dissolution rates at 10°C of approximately 0.0087, 0.0063 and 0.0013 mg/min/cm^2 for TNT, HMX, and RDX, respectively.

Environmental factors such as temperature and pH can affect solubility and therefore chemical fate and toxicity (Huang et al. 2000; Lynch et al. 2001). Based on his studies, Lynch (2002) concluded that the dissolution of

TNT, RDX, and HMX was not affected by pH (between pH 4.2 and 6.2). However, dissolution was significantly affected by temperature over the range tested (3°–33°C), with every 10°C increase in temperature approximately doubling the dissolution rates (Lynch 2002).

C. Volatilization

Because of their low vapor pressures (P_v), volatilization is an insignificant environmental pathway for most common solid-phase TNT, RDX, and HMX. At environmental temperatures (0°–40°C), most energetic compounds are crystalline solids with vapor pressures of the order of 10^{-8} to 10^{-17} atm. As a result, sublimation, the direct mass transfer from the solid to vapor phase, is negligible.

Compounds may also enter the vapor phase via volatilization from the aqueous phase. Henry's law describes the partitioning of a constituent between the vapor and aqueous phases. As a result, the Henry's law constant (K_H) provides an insight into the volatility of a solute. Energetic compounds with K_H values greater than 10^{-5} atm-m^3/mol may volatilize from aqueous solutions; however, the rate of volatilization is influenced by liquid and vapor film resistances. With K_H values of 10^{-7} to 10^{-15} atm-m^3/mol, TNT, TNT transformation products (2,4-DNT, 2,6-DNT, 2-A-4,6-DNT, 4-A-2,6-DNT), RDX, and HMX do not readily volatilize when dissolved in the aqueous phase.

D. Adsorption

The term adsorption refers to a process by which a dissolved chemical (solute) accumulates at an interface. An interface or sorbent may take the form of a mineral, amorphous grain coatings of metal oxyhydroxide, humic material, organic/inorganic colloids, or microorganisms. The degree of partitioning between the solute and the sorbent is dependent on the physicochemical properties of both the solute and the sorbent as well as prevailing environmental conditions. Often partitioning of solute is described by the term sorption, which includes both adsorption and absorption processes. A number of reactions are associated with sorption of a solute to a sorbent; these include hydrophobic partitioning, hydrogen bonding, ion exchange, and chemisorption.

The majority of research conducted on energetic sorption has involved investigation of soil-sorption processes. Numerous studies have shown that TNT can be reversibly sorbed by soil. Hydrogen bonding and ion exchange have been suggested as the possible mechanisms for association between the nitro functional groups and the soil sorbent (Brannon et al. 2002; McGrath 1995; Pennington and Patrick 1990; Xue et al. 1995). Haderlein et al. (1996) showed the variability in TNT sorption onto clay materials with differing exchangeable cations. TNT sorption was four orders of magnitude greater with homoionic K^+ or NH_4^+ clays compared to that when Ca^{2+}, Na^+,

Mg^{2+}, or Al^{3+} were the exchangeable cations. As a result of these findings, Haderlein et al. (1996) suggested that TNT sorption onto soil, sediment, and suspended solids in freshwater environments (where Ca^{2+} dominates) would be significantly lower than that observed in saline environments dominated by K^+ and Na^+. Brannon et al. (2002) also demonstrated the effect of groundwater cation composition on the sorption of TNT. Saturation of aquifer soil cation-exchange sites with K^+ and NH_4^+ resulted in as much as a 9,780% increase in TNT sorption. However, this phenomenon was not observed during studies with RDX and HMX. K^+ or NH_4^+ saturation of aquifer soil cation-exchange sites did not consistently increase sorption of the nitramine energetic compounds.

During sorption–desorption studies with military range soil and RDX, TNT, and 2,4-DNT, Yamamoto et al. (2004) observed linear distribution coefficients (K_d); however, K_d were dependent on the fraction of organic carbon in the soil. TNT and 2,4-DNT were sorbed more strongly than RDX; desorption of 2,4-DNT was negligible whereas the desorption rate of RDX ($1.5\,L\,kg^{-1}$) was considerably slower than its sorption rate ($0.37\,L\,kg^{-1}$). As a result of these findings, Yamamoto et al. (2004) suggested that applying sorption-derived K_d values for transport modeling may significantly overestimate the flux of energetic compounds from the soils studied. In contrast to the study of Yamamoto et al. (2004), Monteil-Rivera et al. (2003) determined that the soil organic carbon content did not significantly affect HMX sorption. Sorption distribution coefficients for HMX were 2.5 and $0.7\,L\,kg^{-1}$ for soil containing 8.4% and 0.33% total organic carbon, respectively.

Studies by Leggett (1985), Ainsworth et al. (1993), Haderlein et al. (1996), and Singh et al. (1998) suggest that sorption of RDX in soil is not a significant attenuation process. However, Sheremata et al. (2001) suggested that although RDX is not extensively sorbed, sorption is nearly irreversible. Although few sorption studies have been performed using HMX, the data indicate that HMX is less sorbed and more mobile than TNT (Price et al. 1998).

Functional groups contained on the energetic compound influences sorption-desorption reactions. In sorption studies using "natural topsoil" and nitroaromatic compounds, Sheremata et al. (1999) determined that the sorption capacity constants for TNT, 4-A-DNT, and 2,4-diamino-nitrotoluene (2,4-DANT) increased with the number of amino groups with significant sorption–desorption hysteresis. However, during sorption studies with illite, K_d values increased with the number of nitro groups. Sheremata et al. (1999) concluded that both the soil type and nitroaromatic compound have significant effects on K_d.

One of the major weaknesses of sorption studies summarized above is the lack of focus on the capacity for sorbing material to release sorbed explosives compound. Given that desorption is the underlying basis for bioavailability, the lack of information on the desorption of such chemicals may lead to both an underestimation of risk and difficulties with the development of effective management strategies.

E. Photolysis

Photolysis is the alteration of a compound caused by direct or indirect effects of light input (Glover and Hoffsomer 1979). Transformation may occur as a result of the direct absorption of light energy, as influenced by the wavelength and intensity, or via the transfer of energy from a photosensitized compound (e.g., peroxide, ozone, humic compounds). Photolysis has been identified as one of the major processes affecting the transformation of energetic compounds in waste streams and surface water bodies (McGrath 1995).

Nitroaromatic energetic compounds, such as TNT, commonly undergo photolysis via oxidation of the methyl group, reduction of nitro groups, or dimer formation. As a result of these reactions, nitrobenzenes, benzaldehydes, azoxydicarboxylic acids, and nitrophenols are produced (Spanggord et al. 1980a). Nitramine compounds, such as RDX and HMX, are also susceptible to photolysis. The resulting transformation products include azoxy compounds, ammonia, formaldehyde, nitrate, nitrite, nitrous oxide, and *N*-nitroso-methylenediamine (Glover and Hoffsommer 1979).

Limited data are available regarding the photolysis half-lives of TNT, RDX, and HMX. Photolysis half-lives have been estimated to range from 0.5 to 22 hr for TNT (Spanggord et al. 1980b; Mabey et al. 1983), from 9 hr to 14 d for RDX (Spanggord et al. 1980b), and from 1.4 to 70 d for HMX (Spanggord et al. 1980b, 1982). Few data are available regarding photolysis half-lives for TNT transformation products (Brannon and Pennington 2002).

F. Hydrolysis

Hydrolysis is a reaction in which a functional group in an organic compound reacts with water to form a new carbon–oxygen bond. Amine, amide, nitrile, and carboxylic acid ester function groups are susceptible to hydrolysis whereas aromatic nitro compounds, aromatic amines, aldehydes, and benzenes are generally resistant. Transformation products resulting from hydrolysis reactions are similar to those produced during photolysis and biotransformation, and as a result differentiating the typical slow hydrolysis process from other reactions in the field is difficult (McGrath 1995; Balakrishnan et al. 2003).

Nitroaromatics and aromatic amines are generally resistant to hydrolysis, although hydrolysis may occur at elevated pH. A number of researchers have demonstrated the alkaline hydrolysis of TNT when the pH of the aqueous or soil system is adjusted to greater than 10 through the addition of NaOH or $Ca(OH)_2$ (Bajpai et al. 2004; Emmrich 1999; Hwang et al. 2006;

Saupe et al. 1998; Thorn et al. 2004). There have been some reports of the hydrolysis of RDX and HMX under alkaline conditions; however, the rate of hydrolysis was extremely slow (Spanggord et al. 1980a, 1982). Balakrishnan et al. (2003) observed the formation of NO_2, HCHO, HCOOH, NH_3, and N_2O following the alkaline hydrolysis of RDX and HMX at $pH > 10$ with pentahydro-3,5-dinitro-1,3,5-triazacyclohex-1-ene being tentatively identified as an intermediate in the hydrolysis reaction. It was proposed that the initial denitration of cyclic nitramines was sufficient to cause ring cleavage and spontaneous decomposition of the energetic compounds (Balakrishnan et al. 2003).

G. Reduction

Energetic compounds containing nitro functional groups are susceptible to abiotic reduction. Nitro groups are reduced through a series of reactions to amino groups, with the process being sensitive to pH and redox potential (McGrath 1995). The rate of reduction in the environment is highly variable, and microbial processes are able to accelerate some reduction reactions. Evidence suggests that abiotic reduction reactions require activation by solid catalysts such as iron compounds, clay minerals or organic macromolecules. Gregory et al. (2004) observed RDX transformation by ferrous iron in aqueous suspensions of magnetite. Sequential reduction resulted in the formation of the nitroso intermediate hexahydro-1-nitroso-3,5-dinitro-1,3,5-triazine (MNX), hexahydro-1,3-dinitroso-nitro-1,3,5-triazine (DNX), and hexahydro-1,3,5-trinitroso-1,3,5-triazine (TNX), and NH_4^+, N_2O, and formaldehyde end-products. In addition, pH was shown to influence the rate of transformation. An increase in pH resulted in greater ferrous iron adsorption, which caused an increase in the rate of RDX transformation (Gregory et al. 2004).

TNT reduction by ferrous iron has also been observed by Hofstetter et al. (1999). Ferrous iron present at the surface of Fe(III)hydroxides and hydroquinone moieties of organic matter in the presence of H_2S were able to abiotically reduce TNT. Although TNT was completely reduced to the corresponding aromatic polyamines, these products were stable and did not undergo further transformation under iron reducing conditions.

Abiotic reduction of energetic compounds may also be achieved through the application of zero valent iron. A number of researchers have demonstrated the potential of zero valent iron for the reduction of TNT and RDX in aqueous or soil matrices (Bandstra et al. 2005; Devlin et al. 1998; Gregory et al. 2004; Park et al. 2004; Singh et al. 1998, 1999). The rate of HMX transformation by zero valent iron is significantly less than that by TNT and RDX; however, the application of cationic surfactants, thereby increasing HMX solubility, can increase transformation rates (Park et al. 2004). However, Park et al. (2004) demonstrated that zero valent iron-mediated transformation of HMX may be inhibited in the presence of RDX.

Although abiotic reduction of energetic compounds may occur under environmental conditions, it is practically impossible to distinguish between biotic and abiotic transformation.

H. Irreversible Surface Reactions

Evidence suggests that irreversible binding of nitroaromatic compounds is a mechanism influencing the fate and transport of these compounds in the environment. A number of mechanisms have been proposed that account for the irreversible binding of energetics to solid surfaces:

- Covalent bonding to specific functional groups in organic and/or mineral solids
- Polymerization and oligomerisation of energetics and organic compounds
- Entrapment of energetics within clay mineral interlamellae or other nanoporous material (McGrath 1995)

Aromatic amines, such as amino-dinitrotoluenes and diamino-nitrotoluenes, may sorb to humic material (carbonyls-aldehyde or ketone moieties) through the formation of an imine. Following this reaction, covalent binding proceeds via the addition of the amine to a quinoidal structure, oxidation to form a nitrogen-substituted quinoid ring, and reaction locking the nitrogen group into a heterocyclic humate structure (Parris 1980). Covalently bound amines are extremely resistant to further reactions or extraction.

I. Biotransformation

Numerous studies have shown the potential of bacteria and fungi to degrade TNT, RDX and HMX under aerobic and anaerobic conditions (Tables 4–6). Degradation may result in mineralization of the compound (i.e., conversion to carbon dioxide, methane, nitrates) or conversion to transformation products (e.g., aromatic amines). The fate and transport of microbial transformation products may then be subject to a number of reactions including polymerization, covalent binding, and complexation. Degradation of energetic compounds may occur as a result of the following mechanisms:

- Utilization of the compound as a sole carbon and energy source
- Cometabolism, a reaction in which microorganisms transform a compound even though the compound cannot serve as an energy source for the organisms; to degrade the compound, the microorganism requires the presence of other compounds, primary substrates, that can support its growth
- Utilization of the compound as a nitrogen source (Boopathy et al. 1998)

The following sections provide an overview of the microbial degradation of TNT, RDX, and HMX under aerobic and anaerobic conditions.

Table 4. Degradation of TNT by Bacteria and Fungi.

Organism	References
Bacteria (aerobic)	
Acinetobacter johnsoni, Acinetobacter junii, Agrobacterium sp. *2PC, Alcaligenes eutrophus, Anabaena* sp., *Arthrobacter globiformis, Arthrobacter* sp. *RP17, Bacillus cereus, B. subtilis, Bacillus* sp., *Corynebacterium glutamicum, Corynebacterium* sp. *Nap2, Cytophaga pectinovora, Flavobacterium odoratum, Klebsiella* sp. *1PC, Klebsiella pnueomoniae, Micrococcus luteus, Mycobacterium* sp. *HL4-NT-1, M. vaccae strain JOB5, Myxococcus xanthus, Pseudomonas aeruginosa, P. aeruginosa MA01, P. cepacia, P. fluorescens, P. pseudoalcaligenes JS52, P. putida, P. putida strain KP-T202, Pseudomonas* sp. *clone A, Pseudomonas* sp. *Toll A, Pseudomonas* sp. *JS150, Pseudomonas* sp. *JLR11, Pseudomonas DFC49, Pseudomonas* sp., *Pseudomonas putida JLR11, Rahnella aquitilis BFB, R. erthropolis, R. globerulus, R. rhodocrouss, Rhizobium* sp. *T10, Rhizobium* sp. *B5, Rhizobium* sp. *M8, Rhodococcus* sp. *TF2, Sphingomonas capsulata, Staphylococcus* sp., *Streptomyces albus, S. chromofuscus A11, S. griseus*	Alvarez et al. (1995), Boopathy and Melancon (2004), Boopathy et al. (1997), Caballero et al. (2005a), Caballero et al. (2005b), Esteve-Núñez and Ramos (1998), Fiorella and Spain (1997), Fuller and Manning (1997), Haïdour and Ramos (1996), Jones et al. (1995), Kalafut et al. (1998), Kim and Song (2005), Kim et al., (2002), Labidi et al. (2001), Park et al. (2002), Park et al. (2003a), Park et al. (2003b), Patsi-Grigsby et al. (1996), Pavlostathis and Jackson (1999), Vanderberg et al. (1995), Vorbeck et al. (1994)
Bacteria (anaerobic)	
Clostridium acetobutylicum, C. bifermentans, C. pasteurianum, Enterobacter cloacae PB2, Escherichia coli, Serratia marcescens, Veillonella alkalscens, Desulfobacterium indolicum, Desulfovibrio sp., *D. gigas, D. desulfuricans, D. vulgaris, Desulfovibrio* sp., *Methanococcus strain B, M. deltae, M. thermolithotrophicus, Methanosarcina barkeri*	Boopathy (1994), Boopathy and Manning (1996), Boopathy et al. (1993), French et al. (1998), Fuller and Manning (1997), Hughes et al. (1998), Kutty and Bennett (2005), Kutty and Bennett (2006), Lewis et al. (1996), McCormick et al. (1976), Montpas et al. (1997), Preuss et al. (1993), Yin et al. (2005a), Yin et al. (2005b)

Table 4. *Continued*

Organism	References
Fungi (aerobic)	
Absidia sp., *Acremonium* sp., *Agaricus aestivalis TMAestl, Agaricus bisporus MWA80-7, Agrocybe praecox TM70.84, Agrocybe praecox YM70.3.1, Alternaria* sp. *TMRZ/WN2, Aspergillus terrus MWi458, Bjerkandera adjusta DSM 3375, Ceratocystis coerulescens, Clitocybe odora TM3, Clitocybula dusenii DSM 11238, Clitocybula dusenii TMB12, Coprinus comatus TM6, Cunninghamella elegans DSM1980, Cyathus stercoreus 36910, Cylindrocarpon* sp., *Fomes fomentarius MWF01–4, Fusarium* sp. *TMS21, Gliocladium* sp., *Heterobasidion annosum TM5P2, Hypoloma fasciculare TM5.2, Irpex lacteus, Kuehneromyces mutabilis TME, Lentinus lepideus, Lepista nebularis TM2, Mucor mucedo DSM810, Nematoloma forwardii DSM 11239, Neurospora crassa TM, Paxillus involutus TM2, Penicillium frequentans ATCC 96048, Penicillium* sp. *DSM 11168, Phanerochaete chrysosporium ATCC 1767, P. chrysosporium ATCC 24725, P. chrysosporium BKM-F-1767, P. chrysosporium DSM 1556, P. chrysosporium, P. sordida HHB-8922, Phlebia brevispora HHB-7030, Phlebia radiata ATCC 64658, Pleurotus ostreatus TMPost, Rhizoctonia solani Mwi5, Stropharia rugoso-annulata DSM11373, Stropharia rugosoannulata, Trametes sueveolens MWT03-2, Trametes versicolor DSM 11269, Trametes versicolor TM5, Trichoderma harzianum, Trichoderma* sp.	Bumpus and Tatarko (1994), Donnelly et al. (1997), Dutta et al. (1998), Eilers et al. (1999), Fernando et al. (1990), Hawari et al. (1999), Hess and Schrader (1998), Hodgson et al. (2000), Kim and Song (2000a), Kim and Song (2000b), Kim and Song (2003), Michels and Gottschalk (1994), Rho et al. (2001), Samson et al. (1998), Scheibner et al. (1997), Spiker et al. (1992), Stahl and Aust (1993), Sublette et al. (1992), Van Aken and Agathos (2001), Van Aken et al. (1997), Van Aken et al. (1999), Weber et al., (2002), Weiss et al. (2004a), Weiss et al. (2004b)

Table 5. Degradation of RDX by Bacteria and Fungi.

Organism	References
Bacteria (aerobic)	
Burkholderia sp., *Methylobacterium* sp. *strain BJ001, Rhizobium rhizogenes, Rhodococcus rhodochrous, Rhodococcus* sp. *strain A, Rhodococcus* sp. *strain D22, Rhodococcus* sp. *strain DN22, Rhodococcus* sp. *strain YH11, Shewanella Halifaxensis, Shewanella sediminis, Stenotrophomonas maltophilia*	Bhushan et al. (2003a), Binks et al. (1995), Coleman and Duxbury (1999), Coleman et al. (1998), Coleman et al. (2002), Jones et al. (1995), Lee and Brodman (2004), Seth-Smith et al. (2002), Tekoah and Abeliovich (1999), van Aken et al. (2004b), Zhao et al. (2005), Zhao et al. (2006)
Bacteria (anaerobic)	
Acetobacterium malicum strain HAAP-1, Acetobacterium paludosum, Bacullus HAW-OC6, Citrobacter freundii, Clostridium acetobutylicum ATCC 824, Clostridium bifermentans, Clostridium butyricum, Clostridium celerecreseens, Clostridium saccharolyticum, Clostridium sp. *strain EDB2, Desulfovibria* spp, *Desulfovibrio desulfuricans, Enterobacter cloacae ATCC 43560, Gordonia sp KTR9, Halomonas HAW-OC4, Klebsiella pneumoniae strain SCZ-1, Marinobacter HAW-OC1, Morgenella morganii, Providencia rettgeri, Pseudoalteromonas HAW-OC2, Pseudoalteromonas HAW-OC5, Williansia* sp. *KTR4*	Adrian and Arnett (2004), Beller (2002), Bhatt et al. (2005), Bhushan et al. (2004), Kitts et al. (1994), McCormick et al. (1981), Pudge et al. (2003), Sherburne et al. (2005), Thompson et al. (2005), Young et al. (1997b), Zhang and Hughes (2003), Zhao et al. (2002), Zhao et al. (2003a), Zhao et al. (2003b)
Fungi (aerobic)	
Phanerochaete chrysosporium, Cladosporium cladosporioides, Phanerochaete chrysospotium, Aspergillus niger	Bhushan et al. (2002), Fernando and Aust (1991), Fournier et al. (2004a), Lee and Brodman (2004), Sheremata and Hawari (2000)
Consortia (anaerobic)	
Municipal sludge, horse manure, domestic sludge, methanogenic mixed culture, soil, groundwater	Adrian et al. (2003), Davis et al. (2004), Hawari (2000), Hawari et al. (2000c), McCormick et al. (1981), Young et al. (1997a), Young et al. (1997b)

Table 6. Degradation of HMX by Bacteria and Fungi.

Organism	References
Bacteria (aerobic)	
Methylobacterium sp.	van Aken et al. (2004a)
Bacteria (anaerobic)	
Clostridium sp. strain EDB2, *Clostridium bifermentans* sp. *strain HAW-1*	Bhushan et al. (2004), Zhao H. S. et al. (2004)
Fungi (aerobic)	
Phanerochaete chrysosporium, Pleurotus ostreatus	Axtell et al. (2000), Fournier et al. (2004b)
Consortia (anaerobic)	
Marine sediment, methanogenic mixed culture, microbial consortium, anaerobic sludge consortia, anaerobic sludge, Thibodaux sewage sludge	Adrian et al. (2003), Bhatt et al. (2005), Boopathy (2001), Groom et al. (2001), Hawari et al. (2000a), Hawari et al. (2000b), Morley et al. (2002), Zhao et al. (2004a)
Other (anaerobic)	
Xanthine oxidase system	Bhushan et al. (2003b)

TNT Degradation

Over the past 25 yr, research has been performed on the bacterial and fungal degradation of TNT (see Table 4). TNT is rapidly transformed by bacteria under aerobic and anaerobic conditions to amino-derivatives (2-ADNT, 4-ADNT, 2,4-DANT, 2,6-DANT) (Fig. 3); however, the nitroaromatic is poorly mineralized. Under strict anaerobic conditions, triaminonitrotoluene (TAT) may also be produced, when all three nitro groups are substituted by NH_2. These products may react further via biotic or abiotic processes to form azo, azoxy, hydrazo, phenolic, and acetyl derivatives.

Fungi have the capacity to degrade TNT via the nonspecific extracellular enzyme systems of lignin peroxidase, manganese peroxidase, and laccase. Unlike bacteria, fungi have the ability to mineralize TNT. Scheibner et al. (1997) and Hodgson et al. (2000) reported over 30% mineralization of TNT by *Clitocybula dusenii* TMb12 and *Phanerochaete chrysosporium*, respectively. Figure 4 illustrates a constructed TNT degradation pathway for *Phanerochaete chrysosporium*. Initial transformation of TNT results in the formation of nitroso-toluene (NsT), which undergoes further transformation to ortho- and para-hydroxylamino-4,6-dinitrotoluene (HADNT) and mono- and diaminotoluenes (ADNT and DANT). A number of azo and azoxy, phenolic, and acylated derivatives may be produced from further transformation of HADNT and DANT; these include azoxy-derivatives (4,4′-6,6′-tetranitro-2,2′-azoxytoluene and 2,2′-6,6′-tetranitro-4,4′-azoxytoluene), azo-derivatives (4,4′-6,6′-tetranitro-2,2′-azotoluene and 2,2′-6,6′-tetranitro-

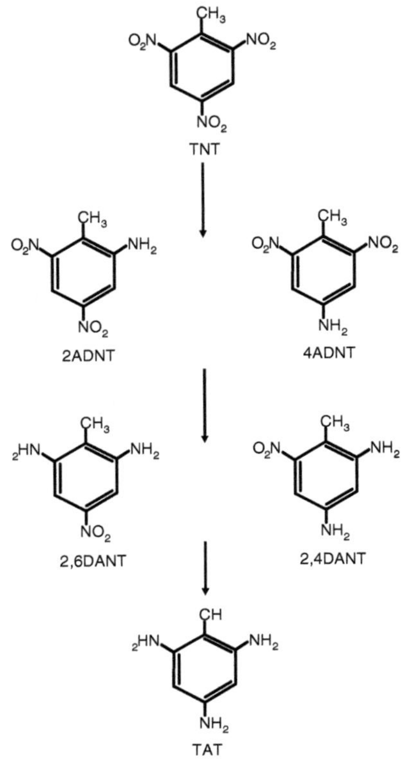

Fig. 3. Bacterial transformation of TNT.

4,4′-azotoluene), hydrazo-derivatives (4,4′-6,6′-tetranitro-2,2′-hydrazotoluene and 2,2′-6,6′-tetranitro-4,4′-hydrazotoluene), and acetylated-derivatives (2-*N*-acetylamino-4,6-dinitrotoluene, 4-*N*-acetylamino-2,6-dinitrotoluene, 2-formylamido-4,6-dinitrotoluene, 4-formylamido-2,6-dinitrotoluene, 4-*N*-acetylamino-2-amino-6-nitrotoluene, 4-*N*-formylamido-2-amino-6-nitrotoluene, 4-*N*-acetylhydroxy-2,6-dinitrotoluene, 4-*N*-acetoxy-2,6-dinitrotoluene, and 4-*N*-acetylamido-2-hydroxylamino-6-nitrotoluene) (Hawari et al. 2000a).

Hawari et al. (1999) suggested that acetylated-derivatives are intermediates in TNT mineralization by *Phanerochaete chrysosporium* as these compounds did not accumulate during TNT degradation. In addition, Pasti-Gribsby et al. (1992) and Spadaro et al. (1992) suggested that azo-derivatives are also amenable for mineralization.

RDX Degradation

Figures 5 and 6 outline biodegradation pathways for RDX under anaerobic conditions. McCormick et al. (1981) were the first to postulate a pathway

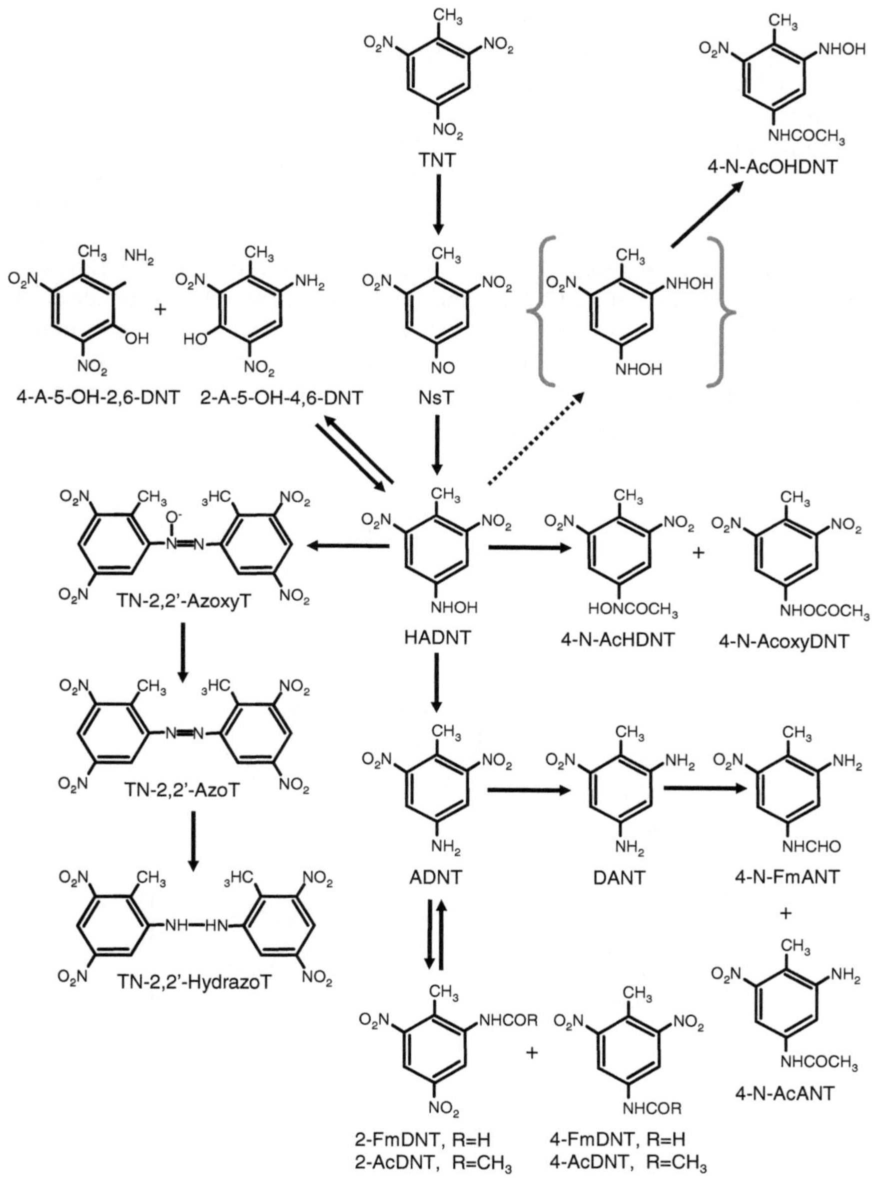

Fig. 4. TNT degradation pathway for *Phanerochaete chrysosporium* (reproduced from Hawari et al. 2000a with permission from Springer). Compounds in brackets represent unidentified products.

for the anaerobic degradation of RDX. Sequential reduction of the nitro groups of RDX resulted in the formation of hexahydro-1-nitroso-3,5-dinitro-1,3,5-triazine (MNX), hexahydro-1,3-dinitroso-nitro-1,3,5-triazine (DNX), and hexahydro-1,3,5-trinitroso-1,3,5-triazine (TNX). Ring cleavage of the nitroso compounds occurred following transformation to hydroxylamine derivatives. Ring cleavage resulted in the formation of hydrazine, 1,1-dimethyl-hydrazines, 1,2-dimethyl-hydrazines, formaldehyde, and methanol (Fig. 5).

An alternative pathway for the anaerobic degradation of RDX was proposed by Hawari et al. (2000c) and further confirmed by Fournier et al. (2002) (Fig. 6). Transformation of RDX was achieved using domestic anaerobic sludge as the inoculum, which resulted in the formation of ring cleavage products methylenedinitramine and bis(hydroxymethyl)nitramine. Both products undergo further degradation to produce nitramine and

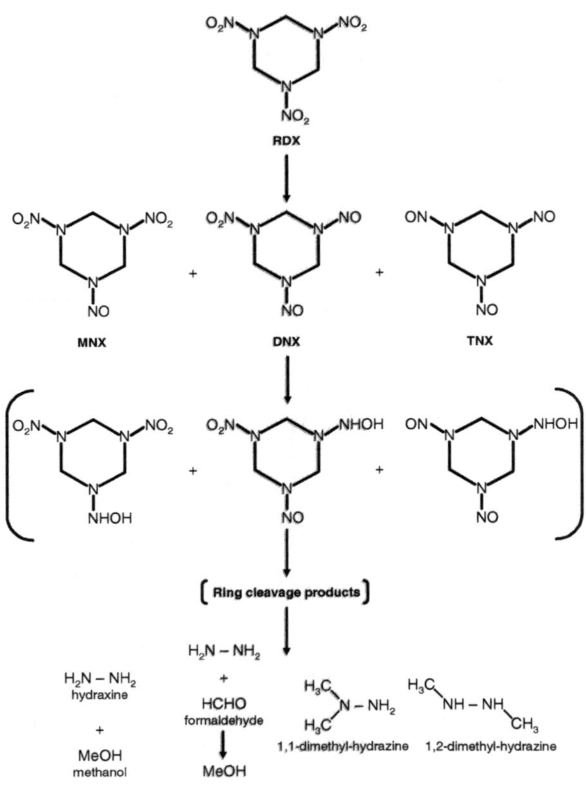

Fig. 5. Metabolic pathway for the anaerobic degradation of RDX: reduction of RDX to nitroso derivatives before ring cleavage and the formation of hydrazines (reproduced from McCormick et al. 1985 with permission from Springer). Compounds in brackets represent unidentified products.

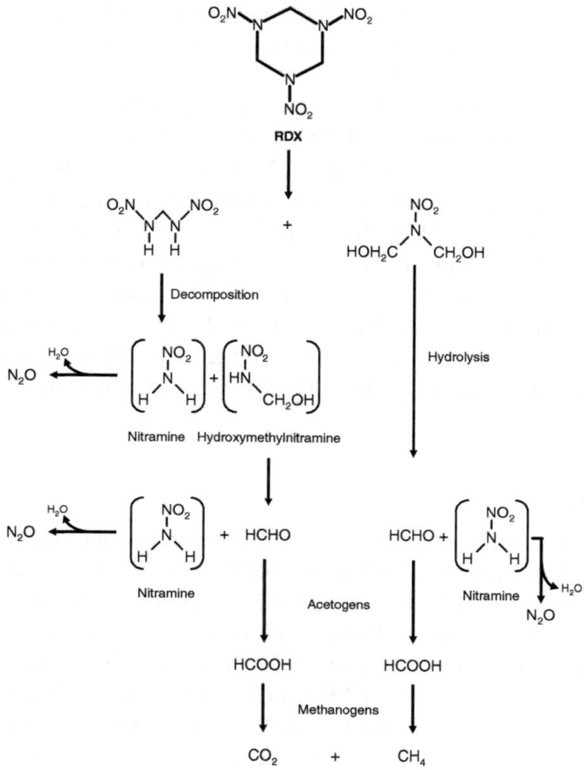

Fig. 6. Direct ring cleavage of RDX during microbial degradation. Ring cleavage resulted in the production of nitrous oxide as the major by-product (reproduced from Hawari et al. 2000a with permission from Springer). Compounds in brackets represent unidentified products.

formaldehyde. Nitramines may spontaneously decompose in water to form nitrous oxide, whereas formaldehyde may be further microbially transformed to carbon dioxide by acetogenic and methanogenic bacteria. Although two alternative pathways for the degradation of RDX have been proposed, enzymes involved in transformation reactions have not been characterized (Adrian and Chow 2001).

RDX may also be degraded under aerobic conditions (Table 5). Studies have shown that RDX can be mineralized by bacteria when the nitramine compound is supplied as a nitrogen source (Binks et al. 1995; Coleman et al. 1998). Although bacterial RDX mineralization has been observed, metabolites associated with compound degradation have not been identified. Fungal mineralization of RDX has also been reported (Fernando and Aust 1991; Sheremata and Hawari 2000). In one study, hexahydro-1-nitroso-3,5-dinitro-1,3,5-triazine (MNX) was detected as a transient degradation product (Sheremata and Hawari 2000); however, MNX did not accumulate in the medium. Nitrous oxide and carbon dioxide were the end products of the mineralisation process.

RDX biodegradation has been demonstrated for contaminated water and marine sediment. Davis et al. (2004) demonstrated the effectiveness of bioremediation for the removal of RDX from contaminated groundwater. After the addition of an alternative carbon source (acetate), RDX was reduced from 100 μg/L to less than 1 μg/L after 27.5 hr. During studies with marine sediment contaminated with energetic compounds, Zhao et al. (2004a) observed significant reduction in RDX concentration when sediments were incubated anaerobically. After 4 d, the concentration of RDX in the aqueous phase (14.7 mg/L) was reduced by 50%; however, carbon supplementation (glucose, acetate, Or citrate) did not enhance the rate of RDX degradation.

HMX Degradation

Far less information is available regarding the microbial degradation pathways for HMX compared to TNT and RDX (see Table 6). The majority of research investigating HMX degradation has involved the use of unidentified microbial communities from anaerobic sediment of sewage sludge (Adrian et al. 2003, Boopathy 2001; Groom et al. 2001; Hawari et al. 2001; Zhao et al. 2004b). Zhao et al. (2004b) observed that HMX degradation by marine sediment, obtained from a military dumping site of UXOs, was enhanced in the presence of glucose, an alternative carbon source. After 50 d, the HMX concentration in the aqueous phase (1.2 mg/L) was reduced by 50%. The disappearance of HMX was accompanied by the formation of a mononitroso derivative.

Adrian et al. (2003) suggested that the addition of hydrogen or electron donors that produce hydrogen may enhance the anaerobic degradation of energetic compounds. In microcosm studies with a methanogenic mixed culture inoculated into a basal salts medium containing HMX (8 μM) and either ethanol, hydrogen, or propylene glycol, 53%, 40%, and 22% of HMX, respectively, was transformed to unidentified products after 29 d. No loss of HMX was observed in microcosms without electron donor supplementation.

Nitroso derivatives and ring cleavage products of HMX were identified during HMX degradation experiments with municipal anaerobic sludge (Figs. 7, 8) (Hawari et al. 2001). Nitroso derivatives were tentatively identified as octahydro-1-nitroso-3,5,7-trinitro-1,3,5,7-tetrazocine, octahydro-1,3-dinitroso-5,7-dinitro-1,3,5,7-tetrazocine, and its isomer octahydro-1,8-dinitroso-3,7-dinitro-1,3,5,7-tetrazocine, whereas ring oxidation products were identified as methylenedinitramine and bis(hydroxymethyl)nitramine. Both nitroso derivatives and ring oxidation products were transient metab-

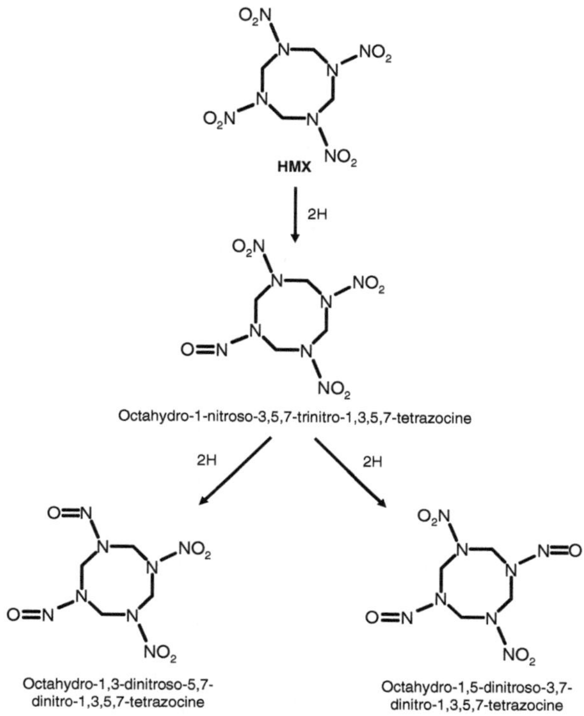

Fig. 7. Proposed pathway for the degradation of HMX data from Kaplan (1993), McCormick et al. (1995), and Hawari et al. (2000).

olites; these compounds disappeared over time to form nitrous oxide and formaldehyde. Further incubation resulted in the biotransformation of formaldehyde to CO_2 (Fig. 8).

HMX may also be degraded under aerobic conditions by bacterial and fungal isolates. Van Aken et al. (2004a) reported the aerobic degradation of HMX by *Methylobacterium* sp. strain BJ001. After 55 d incubation, 61% of HMX was mineralized to CO_2 (approximately 1.5 mg/L). Removal of HMX from soil was reported by Axtell et al. (2000) after amending contaminated soil with a growth substrate and *Pleurotus ostreatus*. During the 62-d treatment period, HMX was reduced from 61 ± 20 mg/kg to 18 ± 7 mg/kg, although transformation products were not observed.

J. Bioaccumulation

A limited number of studies have investigated the bioaccumulation of energetic compounds in plants and animals. Most focused on terrestrial organisms such as earthworms, vegetables, and poplars. Lachance et al.

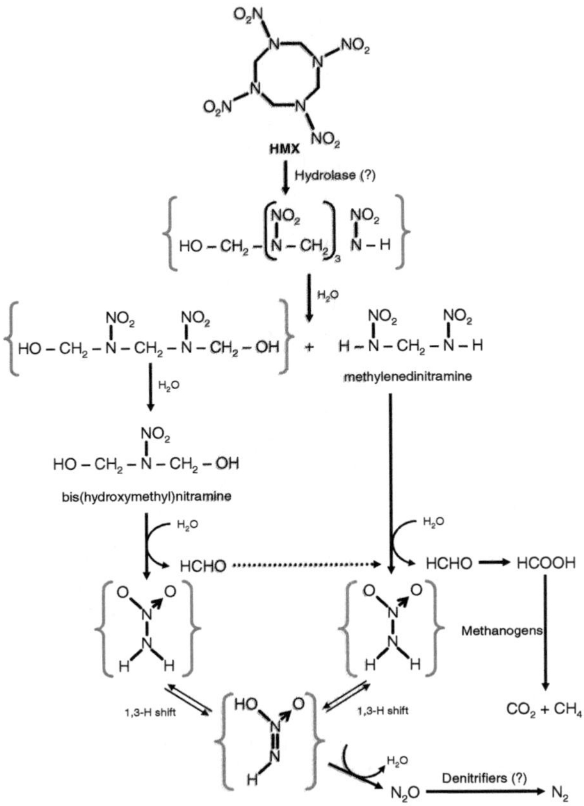

Fig. 8. HMX ring cleavage pathway proposed by Hawari et al. (2000). Reproduced with permission from the American chemical society. Compounds in brackets represent unidentified products.

(2004) investigated the toxicity and bioaccumulation of reduced TNT metabolites (2- and 4-dinitrotoluene, 2,4- and 2,6-diaminonitrotoluene) in the earthworm *Eisenia andrei* exposed to amended forest soil. After 14d exposure, bioaccumulation factors were 5.1, 6.4, 5.1, and 3.2 for 2-amino-4,6-dinitrotoluene, 4-amino-2,6-dinitrotoluene, 2,4-diamino-6-nitrotoluene, and 2,6-diamino-4-nitrotoluene, respectively. The results also suggested that TNT metabolites were as toxic as TNT itself.

The uptake of RDX and HMX by poplar trees (*Populus deltoids* × *nigra*, DN34) has been investigated by a number of researchers (Johnson et al. 1999; Thompson et al. 1998, 1999; Yoon et al. 2002). Energetic uptake was investigated using hydroponic solutions and poplar cuttings. Johnson et al. (1999) observed that up to 60% of the supplied RDX was translocated and accumulated in leaf tissue of the poplar. In HMX uptake studies, 70% of supplied HMX was translocated and accumulated in the leaves (Yoon et al. 2002). In both studies, energetic metabolites were not detected in plant tissue extracts, indicating that RDX and HMX are not transformed during exposure of poplars. In addition, Yoon et al. (2002) determined that the majority of HMX accumulated in fallen poplar leaves could be leached after 5 d, resulting in resolubilized HMX and a significant ecological concern.

In contrast to RDX and HMX, TNT tends to accumulate in the roots of plants with limited transport to other organs. In a study with TNT-spiked soil (10 mg/kg) and bush beans, Cataldo et al. (1989) reported TNT concentrations of <0.6, 9.0, 24.0, and 104 mg/kg in the seeds, leaves, stems, and roots, respectively. Poplar trees may also take up significant concentrations of TNT with most of the contaminant residing in the roots (Thompson et al. 1998).

Belden et al. (2005a) investigated the accumulation of RDX in the channel catfish (*Ictalurus punctatus*) and an oligochaete (*Lumbriculus variegates*). Uptake of RDX by oligochaetes was rapid compared to that by catfish; however, elimination was also more rapid in the oligochaete. As a result, both organisms had similar RDX bioconcentration factors (2.0–2.1 mL/g). When catfish were fed oligochaetes exposed to high concentrations of RDX, accumulation was minimal, indicating that RDX uptake via the aqueous route was the major uptake pathway. In a subsequent study, Belden et al. (2005b) assessed the potential of TNT to accumulate in the aquatic organisms *Chironomus tentans*, *Lumbriculus variegates*, and channel catfish (*Ictalurus punctatus*). TNT bioaccumulation was minimal because of the rapid transformation of the nitroaromatic; however, TNT transformation products accumulated to a greater degree in all three species. Belden et al. (2005b) suggested that further research identifying and determining the relative toxicities of TNT transformation products was warranted to fully evaluate the impact of TNT exposure on aquatic organisms.

IV. Energetics: Toxicity to Ecological Receptors

A. Terrestrial and Aquatic Receptors

Research investigating the ecological effects of energetic compounds has mainly focused on terrestrial and freshwater systems. Data from these studies have been collected from tests utilising a variety of reporter organisms including plants, algae, invertebrates, vertebrates, and microorganisms (Tables 7, 8). To date, the majority of research conducted on ecological effects of energetic compounds has been performed on TNT and its metabolites: few studies have been conducted on RDX and HMX. Table 7 lists test organisms and endpoints for ecological investigation of energetic compounds, and Table 8 shows the range of IC_{50} values determined for

Table 7. Terrestrial, Freshwater, and Marine Organisms Used for Assessing the Ecotoxicity of Energetic Compounds.

Organism	Test	Matrix	References
Microorganisms			
Escherichia coli Indigenous soil microorganisms *Vibrio fischeri*	Microtox, nitrification, dehydrogenase activity, substrate-induced respiration, basal respiration, SOS Chromotest	Antitank firing range soil, spiked soil, manufacturing facility soil	Dodard et al. (1999), Hawari et al. (2000b), Hawthorn et al. (2000), Robidoux et al. (2004b), Sunahara et al. (1998, 1999), Zeng et al. (2004)
Algae			
Selenastrum capricornutum *Ulva fasciata*	Growth inhibition algal zoospore germination, germling growth	Antitank firing range soil, spiked soil, manufacturing facility soil, spiked water	Dodard et al. (1999), Nipper et al. (2001), Robidoux et al. (2004b), Sunahara et al. (1999)
Invertebrates			
Ampelisca abdita *Aporrectodea rosea* *Arbacia punctulata* *Chironomus tentans* *Dinophilus gyrociliatus* *Eisenia andrei* *Enchytraeus albidus* *Enchytraeus crypticus* *Eohaustorius estuarius* *Folsomia candida* *Hyalella azteca* *Lumbricus terrestris* *Mysidopsis bahia* *Tubifex tubifex*	Survival, lysosomal membrane stability, reproduction test, avoidance response test, uptake, fertilization, embryological development, survival	Antitank firing range soil, spiked soil, sediment, and water	Conder et al. (2004a 2004b), Dodard et al. (2003, 2004, 2005), Jarvis et al. (1998), Kuperman et al. (2003), Kuperman et al. (2005), Lachance et al. (2004), Lotufo and Farrar (2005), Nipper et al. (2001), Renoux et al. (2000), Robidoux et al. (1999, 2000, 2002a, 2002b, 2004a, 2004b, 2005), Rosen and Lotufo (2005), Schaefer (2004), Schafer and Achazi (1999), Simini et al. (2003), Steevens et al. (2002)

Table 7. *Continued*

Organism	Test	Matrix	References
Vertebrates			
Ambystoma tigrinum *Colinus virginianus* *Cyprinodon variegates* *Danio rerio* Fischer 344 rats *Ictalurus punctalus* *Menidia beryllina* *Oncorhynchus mykiss* *Peromyscus maniculatus* *Pimephales promelas* *Plethodon cinereus* *Sciaenops ocellatus* Sprague–Dawley rats	Survival, weight gain, organ to body weight ratio, splenic phagocytic cell function, larvae survival	Spiked soil and water 2004, 2005),	Ek et al. (2005), Johnson et al. (1999, 2000, Levine et al. (1981), Lotufo and Lydy (2005), Lotufo et al. (2001), Meyer et al. (2005), Mukhi et al. (2005), Nipper et al. (2001), Ownby et al. (2005), Smith et al. (2006)
Plants			
Allium cepa *Arabidopsis thaliana* *Brassica rapa* Metzg *Bromus inermis* Leyss. *Cucumis sativus* *Echinochloa crusgalli* L. *Festuca arundinacea* *Hordeum vulgare* *Lactuca sativa* L. *Lens culinaris* Med. *Lolium perenne* L. *Lepidium sativum* L. *Medicago sativa* L. *Panicum virgatum* L. *Raphamus sativus* *Sinapis alba* L. *Triticum aestirum* L.	Root elongation, seedling emergence, fresh shoot biomass	Water, sand/soil mixture, antitank firing range soil	Gong et al. (1999), Kim et al. (2004), Peterson et al. (1996), Peterson et al. (1998), Picka and Friedl (2004), Price et al. (2002); Rocheleau et al. (2006), Robidoux et al. (2003), Robidoux et al. (2004b), Scheidemann et al. (1998), Simini et al. (1995)

Table 8. Toxicity of Energetic-Spiked Water, Seawater, Porewater, Soils, and Sediments to Ecological Receptors.

Organism	Test	Matrix	IC$_{50}$ (ppm)									References
			TNT	2,4-DNT	2,6-DNT	2-ADNT	4-ADNT	2,4-DANT	2,6-DANT	RDX	HMX	
Microorganisms												
Vibrio fischeri	Microtox (15 min)	Spiked water, extracts from spiked soils	0.3–1.5	36–73	2.5–3.6	>22	>26	53–81	>51; >101	>42	>6.5	Dodard et al. (1999), Sunahara et al. (1998, 1999)
Vibrio fischeri	Microtox (30 min)	Spiked water, extracts from spiked soils	2.0–3.6			21–230	10–122	48	>100			Drzyzga et al. (1995), Frische (2002), Johnson et al. (1994)
Algae												
Selanastrum capricornutum	Growth inhibition (96 hr)	Extracts from spiked soils	0.5–1.0			1.9–3.4	10–13	23–96	29–43			Sunahara et al. (1998, 1999)
Ulva fasciata	Germination	Spiked seawater, spiked pore water, pore water from spiked sediment	2.5	2.5	0.092–17					12		NFESC (2000a) NFESC (2000b), Carr and Nipper (2003)

Table 8. *Continued*

Organism	Test	Matrix	IC_{50} (ppm)									References
			TNT	2,4-DNT	2,6-DNT	2-ADNT	4-ADNT	2,4-DANT	2,6-DANT	RDX	HMX	
Ulva fasciata	Germling length	Spiked seawater, spiked pore water, pore water from spiked sediment	0.76	1.7	<0.087–8					8.1		NFESC (2000a) NFESC (2000b), Carr and Nipper (2003)
Ulva fasciata	Germling cell number	Spiked seawater, spiked pore water, pore water from spiked sediment	1.4	2.1	<0.087–9					9.8		NFESC (2000a) NFESC (2000b), Carr and Nipper (2003)
Invertebrates												
Arbacia punctulata	Fertilisation	Spiked seawater	>103	68	>84					>75		NFESC (2000a)
Arbacia punctulata	Embryo development	Spiked seawater, spiked pore water	12	51	<0.029–36.9					>75		NFESC (2000a), NFESC (2000b)
Copepod	Survival	Pore water from spiked sediment			49–62							Carr and Nipper (2003)
	Embryo and nauplii survival	Pore water from spiked sediment			60–69							Carr and Nipper (2003)

Table 8. *Continued*

			IC_{50} (ppm)									
Organism	Test	Matrix	TNT	2,4-DNT	2,6-DNT	2-ADNT	4-ADNT	2,4-DANT	2,6-DANT	RDX	HMX	References
Dinophilus gyrociliatus	Survival	Spiked seawater, spiked pore water	7.7	21	0.046–21.1					>49		NFESC (2000a), NFESC (2000b)
Dinophilus gyrociliatus	Laid eggs/female	Spiked seawater	1.8	5.7	2.1					26		NFESC (2000a), NFESC (2000b)
Eisenia andrei	Survival (7 d)	Spiked OECD artificial soil Spiked forest soil	204–400									Robidoux et al. (1999)
Eisenia andrei	Survival (14 d)	Spiked OECD artificial soil Spiked forest soil	132–222			201–228	99–111					Lachance et al. (2004), Renoux et al. (2000), Robidoux et al. (2000)
Eisenia fetida	Survival (14 d)	Spiked forest soil	325									Phillips et al. (1993)

Table 8. *Continued*

Organism	Test	Matrix	IC$_{50}$ (ppm) TNT	2,4-DNT	2,6-DNT	2-ADNT	4-ADNT	2,4-DANT	2,6-DANT	RDX	HMX	References
Enchytraeus albidus	Survival (21 d)	Spiked artificial soil	422 ± 63									Dodard et al. (2003)
	Fecundity (42 d)	Spiked artificial soil	111 ± 34									Dodard et al. (2003)
Enchytraeus crypticus	Survival (7 d)	Spiked Lufa 2.2 soil	570									Schafer and Achazi (1999)
Eohaustorius estuarius	Survival (10 d)	Spiked sediment	28–36									Rosen and Lotufo (2005)
Folsomia candida	Survival (7 d)	Spiked Lufa 2.2 soil	185									Schafer and Achazi (1999)
Hyalella azteca	Survival (10 d)	Sediments	30					3460				
Mysidopsis bahia	Juvenile survival	Spiked seawater	0.98	5.4	5.6					>47		
Vertebtates												
Cyprinodon variegatus	Survival	Spiked seawater	2.3			8.6		>51		9.8–9.9	>aqueous solubility	Gensemer et al. (2004), Lotufo et al. (2001)
Menidia beryllina	Survival	Spiked seawater								7.1		Gensemer et al. (2004)
Sciaenops ocellatus	Larvae survival	Spiked seawater	8.2	48	34					>68		NFESC (2000a)
Xenopus laevis	FETAX assay (96 hr)	Spiked water	3.8			33	23					Saka (2004)

various ecological receptors using TNT, TNT transformation products, RDX, and HMX.

Microbial assays utilising the aquatic bacterium *Vibrio fischeri* (Microtox test) have been conducted to determine the toxicity of energetic compounds in pure systems and contaminated and spiked soils (see Table 8). An advantage of using assays such as the Microtox test is that they are simple, rapid, and relatively cost-effective. Using an acetonitrile-sonication extraction method associated with the Microtox test, Sunahara et al. (1998) determined that TNT was toxic to *Vibrio fischeri* (IC_{50}, 15 min: 0.95 mg/L) whereas RDX was less toxic (IC_{50}, 15 min: 40 mg/L), and HMX was not toxic up to its aqueous solubility. In a subsequent study, Sunahara et al. (1999) determined that soil factors significantly influenced TNT extractability from spiked soils and subsequent toxicity measurement. IC_{50} values for TNT ranged from 0.27 to 0.94 mg/L depending on the soil moisture (20% versus dry) and texture (sandy versus clayey-sandy). These studies suggest that soil significantly influences the bioavailability of TNT and therefore has implications for leachability, degradation, and toxicological processes.

In an attempt to determine whether degradation of TNT and subsequent microbial reduction was associated with a detoxification process, Dodard et al. (1999) tested the toxicity of TNT transformation products using the Microtox test. Results indicated that 2,6-DNT was more toxic than 2,4-DNT whereas reduced metabolites of 2,6-DNT were less toxic than the parent compound (see Table 8). However, partially reduced metabolites of 2,4-DNT were more toxic than the parent compound, raising doubts over the hypothesis that the reductive metabolism of nitroaromatics is associated with a subsequent detoxification process (Dodard et al. 1999). In addition to the Microtox test, other microbial assays have been used for determining the toxicity of energetic compounds in environmental matrices. Viable counts, enzyme activity, substrate-induced respiration, and basal respiration have been used to assess the toxicity of TNT in freshwater (Zeng et al. 2004) and contaminated soils from antitank firing ranges (Robidoux et al. 2004a).

The freshwater unicellular green algae (*Selenastrum capricornutum*) 96-hr growth inhibition test is another popular ecotoxicity assay for assessing the toxicity of energetic compounds. TNT is toxic to *S. capricornutum* with EC_{50} 96-hr values of 0.93 ± 10.22 mg/L (Sunahara et al. 1998). Similar to Microtox assays, toxicity followed the order TNT > RDX >> HMX; in fact, HMX was not toxic at concentrations up to its aqueous solubility. In contrast to Microtox assays, Dodard et al. (1999) determined the relative toxicities of 2,4-DNT and its metabolites to be 2,4-DNT > 2,4-DAT = 4A-2NT = 2A-4NT, indicating that the parent compound was the most toxic.

The effect of energetic compounds on the mortality and reproduction of terrestrial invertebrates has been tested using earthworm, enchytraeid, and collembola bioassays (see Tables 7, 8). The majority of invertebrate tests have been conducted with TNT-spiked soils, but a few studies have investigated the toxicity of firing range soil (containing HMX) using these organisms (Robidoux et al. 2004a,b). In addition to the standard acute and reproduction tests, an avoidance test has been used for assessing ecological impact of energetic-contaminated soils. Schaefer (2004) showed significant repellent effects at a TNT concentration of 29 mg/kg; a concentration of 1,142 mg/kg was required to show toxic effects for acute and reproduction tests. In bioassays conducted by Dodard et al. (2003) using the white potworm (*Enchytraeus albidus*) and TNT-spiked soil, TNT was 5–10 times more lethal to juveniles than adult potworms, whereas Lachance et al. (2004) determined the toxicity of TNT and its metabolites to *Eisenia andrei* to follow the order 4-ADNT > TNT EN>2-ADNT. In collembola assays with *Folsomia candida* and TNT-spiked soil, LC_{50} values for 7-d and 28-d mortality tests were 185 and 110 mg TNT/kg, respectively.

In studies conducted by Robidoux et al. (2001), the chronic toxicity of HMX in artificial spiked soil was determined using the earthworm reproduction test and *Eisenia andrei*. Fecundity was assessed based on total and hatched numbers of cocoons, number of juveniles, and their biomass. Results from these studies indicated that HMX at concentrations as low as 280 ± 12 mg HMX/kg (LOEC) had significant sublethal effects on *Eisenia andrei*.

A dearth of information exists on the phytotoxicity of energetic compounds. The majority of studies utilized agricultural species (e.g., wheat, mustard, lettuce, lentil, barley, cress, turnip, onion, radish, cucumber, alfalfa) (see Table 7) using root elongation, seedling emergence, and biomass production as toxicological endpoints. Some studies assessed the toxicity of TNT to plants under hydroponic conditions (Kim et al. 2004); however, these results are difficult to translate to contaminated soil matrices because of contaminant bioavailability issues. Gong et al. (1999) demonstrated that the phytotoxic response of plants to TNT-spiked soil was species specific. The LOEC for cress and turnip was 50 mg TNT/kg; oat could tolerate TNT concentrations up to 1,600 mg TNT/kg. In other studies, Scheidemann et al. (1998) demonstrated that alfalfa was unable to grow in soil containing 100 mg TNT/kg although bush pea grew at concentrations up to 500 mg TNT/kg. In studies using TNT-spiked artificial soil, Robidoux et al. (2003) observed significant effects on lettuce and barley species. For lettuce, emergence was reduced by 18% (after 5 d) compared to the control group at a TNT concentration of 1,040 mg/kg. For barley, the effects on shoot growth (after 14 d) were significant at TNT concentrations above 56 mg TNT/kg: fresh shoot and fresh root biomass were reduced by 43%–63% and 90%–99%, respectively. Exposure to TNT caused significant effects on lettuce and barley, but no effect was observed in artificial soil spiked with HMX at concentrations up to 1,866 ± 438 mg/kg (Robidoux et al. 2003). Similarly,

Best et al. (2006) observed that biomass of *Lolium perenne* was not influenced by soil RDX concentrations up to 1,540 mg/kg (the highest concentration used) whereas a significant increase in *Medicago sativa* biomass was observed.

In studies using perennial grasses, Peterson et al. (1996, 1998) demonstrated the influence of TNT on seed germination and seedling development. For tall fescue and bromegrass, germination decreased linearly as TNT concentrations increased to 60 mg TNT/L, whereas switchgrass germination was unaffected by TNT at these concentrations. Effects on seedling growth and development, as evidenced by abnormal radical tissue development and reduced secondary root and shoot growth were observed at TNT concentrations above 7.5 mg TNT/L for switchgrass and bromegrass and 30 mg TNT/L for tall fescue (Peterson et al. 1996, 1998).

Plants also have the ability to metabolise energetic compounds via reduction and oxidation processes (Best et al. 2005; Hannink et al. 2002; Robidoux et al. 2003). The distribution of transformation products, conjugates, and bound residues (Bhadra et al. 1999; Vila et al. 2005) appears to be consistent with the green liver concept (Best et al. 2005). Understanding the processes associated with plant uptake and metabolism of TNT, RDX, and HMX is critical for assessing the potential of phytoremediation technologies (Li et al. 2005). Although not part of this review, biochemical studies enabling the degradation of explosives by different plant species are described in reviews by Hannink et al. (2002) and Best et al. (2005).

Vertebrate toxicity tests are useful to assess acute, subacute, and chronic toxicity of energetic compounds by examining the growth, survival, reproductive success, and body burden of test organisms. A number of test organisms have been used to determine the toxicity of energetic compounds, including fish (*Pimephales promelas*, *Oncorhynchus mykiss*, *Danio rerio*) (Burton et al. 1994; Ek et al. 2005; Mukhi et al. 2005), rodents (Dilley et al. 1982; Levine et al. 1981; Reddy et al. 2000), amphibians (*Ambystoma tigrinum, Plethodon cinereus*) (Johnson et al. 1999, 2000, 2004), dogs (Dilley et al. 1982; Levine et al. 1990; Yinon 1990), cats (Yinon 1990), rabbits (Yinon 1990), and guinea pigs (Yinon 1990). Initial studies utilized an animal model (dog) for determining the oral chronic reference dose for human health risk assessment (U.S. Department of Defense 1983); however, recent work has focused on the ecological impact of energetic compounds on terrestrial and freshwater receptors (Burton et al. 1994; Johnson et al. 1999, 2000, 2004; Mukhi et al. 2005).

B. Marine Receptors

A limited amount of research has been performed to develop a marine toxicity database for energetic compounds (NFESC 2000a; Nipper et al. 2001). Nipper et al. (2001) assessed a number of energetic compounds including 2,4-dinitrotoluene, 2,6-dinitrotoluene, TNT, and RDX for their toxicological effects on five marine species including sea urchin (*Arbacia punctulata*), polychaetes (*Dinophilus gyrociliatus*), a macroalga (*Ulva fasciata*), redfish (*Sciaenops ocellatus*), and opossum shrimp (*Mysidopsis bahia*). Various toxicological endpoints were recorded including fertilization toxicity, embryological development, larval and juvenile survival, zoospore germination, and germling growth (see Tables 7, 8). Among the test methods and endpoints, toxicity differed considerably for the energetic compounds tested. The reproductive endpoint (number of eggs per female) for the polychaete was more sensitive than the survival endpoint, whereas germling length was the most sensitive endpoint for the macro algae toxicity tests. For the nitroaromatic compounds, toxicity tended to increase with the level of nitrogenation; i.e., TNT was more toxic than DNT. RDX was the least toxic energetic compound tested (toxicity was not observed for five of the nine test endpoints); in some tests, toxicity was only seen after enhancing RDX's aqueous solubility through the addition of a carrier solvent (e.g., methanol).

In a study by Carr and Nipper (2003), the toxicity of 2,6-DNT and its transformation products was assessed using the macroalgae (*Ulva fasciata*) zoospore germination and germling growth tests. In addition, toxicity tests were performed using the copepod *Schizopera knabeni*. Endpoints were female survival and nauplii hatching and survival. Initially, experiments used filtered seawater spiked with energetic compounds or their transformation products. Neither 2,6-DNT nor its major transformation product, 2-amino-6-nitrotoluene (2-A-6-NT), was highly toxic during the copepod adult female survival test. NOEC values of 277 and 240 μmoles/L were determined for 2,6-DNT and 2-A-6-NT, respectively. However, higher toxicity values were observed during the nauplii hatching rate assay: EC_{50} values of 52 and 242 μmoles/L were calculated for 2,6-DNT and 2-A-6-NT, respectively (Carr and Nipper 2003). In contrast to this, results obtained for the macroalgae toxicity tests demonstrated that the parent compound was more toxic than the transformation product. EC_{50} values for 2-A-6-NT were >160, 123 and 137 μmoles/L for germination, germling length, and cell numbers, respectively, compared to 2,6-DNT EC_{50} values of 73, 20, and 24 μmoles/L (Carr and Nipper 2003).

In a subsequent study, the toxicity of marine pore waters (Puget Sound and Redfin Bay, WA, USA) spiked with energetic compounds, including 2,6-DNT, was assessed using the sea urchin embryological development test, the macroalgae zoospore germination and germling growth tests, and the polychaete survival and reproduction tests (NFESC 2000b). In addition, toxicological assessment of spiked sediments (Puget Sound and Redfin Bay) was assessed using the 10-d acute benthic amphipod (*Ampelisca abdita*) toxicity test. This test involves exposure of *Ampelisca abdita* to marine sediment and determination of animal survival compared to animals exposed to a reference sediment.

Nearly all Puget Sound pore water samples spiked with energetic compounds were more toxic than Redfin Bay samples. In addition, in some cases, pore water toxicity was greater than in filtered seawater spiked with energetic compounds, suggesting that either degradation products or other constituents in pore water were responsible for the exhibited toxicity (NFESC 2000b). The authors suggested that unidentified transformation products may be responsible for the increased toxicity, especially in the urchin embryo assay, and concluded that the sedimentary origin of the spiked pore water has a significant influence on the variability in toxicity test sensitivity (NFESC 2000b).

The fate and dynamics of energetic compounds in sediments will influence their impact on ecological receptors. To illustrate this point, Carr and Nipper (2003) assessed the toxicity of 2,6-DNT in spiked sediment and pore water over a time course period. Uncontaminated sediments [a fine-grained sediment with a total organic carbon (TOC) of 1.1% and a sandy sediment with a TOC of 0.25%] were spiked with the energetic compound, up to 1,500 μmoles/L sediment d.wt., and aged up to 6 mon at 10°C and 20°C. Contaminant transformation and pore water toxicity were assessed during the 6-mon period.

Transformation of 2,6-DNT occurred in all spiked sediments; however, the rate of transformation was greater at 20°C and in the fine-grained sediment. Transformation products identified for 2,6-DNT included 2-A-6-NB (major), 2-nitrotoluene, *n,n*-dimethyl-3-nitroaniline, benzene nitrile, methylamino-2-nitrosophenol, and diaminophenol. Carr and Nipper (2003) proposed that both biotic and abiotic processes were responsible for the transformation of 2,6-DNT. Correspondingly, 2,6-DNT pore water toxicity fluctuated over the time course experiments, varying with the concentration of the parent compound and transformation products. EC_{50} values were calculated at each time point based on the concentration of the parent compound and transformation product at the time of sampling. When contaminants were undetected, EC_{50} values were expressed as percentage pore water.

In general, the toxicity of 2,6-DNT-spiked pore water samples decreased over time resulting in the inability to calculate EC_{50} values for copepod survival, macroalgae germling length, and macroalgae cell numbers after 180 d of pore water aging. However, a slight increase in toxicity was observed after 7 d in the copepod embryo and nauplii survival test. EC_{50} values decreased from 377 and 327 μmoles/L to 178 and 204 μmoles/L in porewater obtained from 2,6-DNT-spiked sandy and fine sediment, respectively. Presumably, unidentified 2,6-DNT transformation products induced a greater toxic response; however, further transformation of these compounds reduced pore water toxicity at the end of the experimental period (180 d) (Carr and Nipper 2003). Carr and Nipper (2003) concluded from these studies that it would be prudent to include some energetic transformation products in the list of analytes of concern during field assessment of potentially contaminated sediments.

C. Ecotoxicity and Bioavailability

The varying IC_{50} values for soil- and sediment-associated vertebrates and invertebrates (see Table 8) reflect the different toxicities of the energetic compounds and the sensitivities of the test organisms, as well as the bioavailability of TNT, RDX, and HMX in contaminated soils and sediments. Bioavailability processes, including physical, chemical, and biological interactions, may significantly influence the exposure of test organisms to soil- and sediment-associated contaminants. Adsorption of energetic compounds to soil and sediment surfaces through hydrophobic partitioning, hydrogen bonding, or chemisorption will influence contaminant bioavailability. For organisms that feed directly on soil and sediment or for those that absorb contaminants across external membranes, association and dissociation of the contaminant with the solid phase plays a vital role in determining contaminant bioavailability and corresponding toxicity responses. Although data are starting to emerge on the ecotoxicological impact of energetic compounds in field soils, a more basic understanding of energetic bioavailability and toxicological responses in aged soil covering a range of soil types is required.

Summary

An explosive or energetic compound is a chemical material that, under the influence of thermal or chemical shock, decomposes rapidly with the evolution of large amounts of heat and gas. Numerous compounds and compositions may be classified as energetic compounds; however, secondary explosives, such as TNT, RDX, and HMX pose the largest potential concern to the environment because they are produced and used in defense in the greatest quantities. The environmental fate and potential hazard of energetic compounds in the environment is affected by a number of physical, chemical, and biological processes. Energetic compounds may undergo transformation through biotic or abiotic degradation. Numerous organisms have been isolated with the ability to degrade/transform energetic compounds as a sole carbon source, sole nitrogen source, or through cometabolic processes under aerobic or anaerobic conditions. Abiotic processes that lead to the transformation of energetic compounds include photolysis, hydrolysis, and reduction. The products of these reactions may be further transformed by microorganisms or may bind to soil/sediment surfaces through covalent binding or polymerization and oligomerization reactions. Although considerable research has been performed on the fate and dynamics of energetic compounds in the environment, data are still gathering on the impact of TNT, RDX, and HMX on ecological receptors. There is an urgent need to address this issue and to direct future research on expanding our knowledge on the ecological impact of energetic transformation products. In addition, it is important that energetic research considers the concept of bioavailability, including factors influencing soil/sediment aging, desorption of energetic compounds from varying soil and sediment types, methods for modeling/predicting energetic bioavailability, development of biomarkers of energetic exposure or effect, and the impact of bioavailability on ecological risk assessment.

Acknowledgments

The project was funded by the Environment, Heritage and Risk Branch of the Australian Department of Defence. The authors would like to acknowledge the support of the Centre for Environmental Risk Assessment and Remediation (CERAR, University of South Australia) and the Cooperative Research Centre for Contamination Assessment and Remediation of the Environment (CRC CARE).

References

Adrian NR, Arnett CM (2004) Anaerobic biodegradation of hexahydro-1,3,5-trinitro-1,3,5-triazine (RDX) by *Acetobacterium malicum* strain HAAP-1 isolated from a methanogenic mixed culture. Curr Microbiol 48:332–340.

Adrian NR, Chow T (2001) Identification of hydroxylamino-dinitroso-1,3,5-triazine as a transient intermediate formed during the anaerobic biodegradation of hexahydro-1,3,5-trinitro-1,3,5-triazine. Environ Toxicol Chem 20:1874–1877.

Adrian NR, Arnett CM, Hickey RF (2003) Stimulating the anaerobic biodegradation of explosives by the addition of hydrogen or electron donors that produce hydrogen. Water Res 37:3499–3507.

Ainsworth CC, Harvey SD, Szecsodt JE, Simmons MA, Cullinan VI, Resch CT, Mong GH (1993) Relationship between the leachability characteristics of unique energetic compounds and soil properties. Project 91PP1800. U.S Army Biomedical Research and Development Laboratory, Fort Detrick, MD.

Alvarez MA, Kitts CL, Botsford JL, Unkefer P (1995) *Pseudomonas aeruginosa* strain MA01 aerobically metabolises the aminodinitrotoluenes produced by 2,4,6-trinitrotoluene nitro-group reduction. Can J Microbiol 41:984–991.

Ampleman G, Faucher D, Thiboutot S, Hawari J, Monteil-Rivera F (2004) Evaluation of underwater contamination by explosives and metals at Point Amour, Labrador and in the Halifax Harbour area. DRDC Valcartier TR 2004-125. Defence Research and Development, Toronto, Canada.

Army, U.S. Department of Defense (1994) Military Explosives TM 9-1300-214. U.S Department of the Army, Washington D.C.

Axtell C, Johnston CG, Bumpus JA (2000) Bioremediation of soil contaminated with explosives at the Naval Weapons Station Yorktown. Soil Sed Contam 9:537–548.

Bajpai R, Parekh D, Herrmann S, Popovic M, Paca J, Qasim M (2004) A kinetic model of aqueous-phase alkali hydrolysis of 2,4,6-trinitrotoluene. J Hazard Mater 106:37–44.

Balakrishnan VK, Halasz A, Hawari J (2003) Alkaline hydrolysis of the cyclic nitramine explosives RDX, HIVIX, and CL-20: new insights into degradation pathways obtained by the observation of novel intermediates. Environ Sci Technol 37:1838–1843.

Bandstra JZ, Miehr R, Johnson RL, Tratnyek PG (2005) Reduction of 2,4,6-trinitrotoluene by iron metal: kinetic controls on product distributions in batch experiments. Environ Sci Technol 39:230–238.

Belden JB, Lotufo GR, Lydy MJ (2005a) Accumulation of hexahydro-1,3,5-trinitro-1,3,5-triazine in channel catfish (*Ictalurus punctatus*) and aquatic oligochaetes (*Lumbriculus variegatus*). Environ Toxicol Chem 24:1962–1967.

Belden JB, Ownby DR, Lotufo GR, Lydy MJ (2005b) Accumulation of trinitrotoluene (TNT) in aquatic organisms. Part 2: Bioconcentration in aquatic invertebrates and potential for trophic transfer to channel catfish (*Ictalurus punctatus*). Chemosphere 58:1161–1168.

Beller HR (2002) Anaerobic biotransformation of RDX (hexahydro-1,3,5-trinitro-1,3,5-triazine) by aquifer bacteria using hydrogen as the sole electron donor. Water Res 36:2533–2540.

Best EPH, Kvesitadze G, Khatisashvili G, Sadunishvili T (2005) Plant processes important for the transformation and degradation of explosives contaminants. Z Naturforsch Sect C J Biosci 60:340–348.

Best EPH, Geter KN, Tatem HE, Lane BK (2006) Effects, transfer, and fate of RDX from aged soil in plants and worms. Chemosphere 62:616–625.

Bhadra R, Wayment DG, Hughes JB, Shanks JV (1999) Confirmation of conjugation processes during TNT metabolism by axenic plant roots. Environ Sci Technol 33:446–452.

Bhatt M, Zhao JS, Monteil-Rivera F, Hawari M (2005) Biodegradation of cyclic nitramines by tropical marine sediment bacteria. J Ind Microbiol Biotechnol 32:261–267.

Bhushan B, Halasz A, Spain J, Thiboutot S, Ampleman G, Hawari J (2002) Biotransformation of hexahydro-1,3,5-trinitro-1,3,5-triazine catalyzed by a NAD(P)H: nitrate oxidoreductase from *Aspergillus niger*. Environ Sci Technol 36:3104–3108.

Bhushan B, Trott S, Spain JC, Halasz A, Paquet L, Hawari J (2003a) Biotransformation of hexahydro-1,3,5-trinitro-1,3,5-triazine (RDX) by a rabbit liver cytochrome p450: insight into the mechanism of RDX biodegradation by *Rhodococcus* sp. strain DN22. Appl Environ Microbiol 69:1347–1351.

Bhushan B, Paquet L, Halasz A, Spain JC, Hawari J (2003b) Mechanism of xanthine oxidase catalyzed biotransformation of HMX under anaerobic conditions. Biochem Biophys Res Commun 306:509–515.

Bhushan B, Halasz A, Thiboutot S, Ampleman G, Hawari J (2004) Chemotaxis-mediated biodegradation of cyclic nitramine explosives RDX9 HMX, and CL-20 by *Clostridium* sp. EDB2. Biochem Biophys Res Commun 316:816–821.

Binks PR, Nicklin S, Bruce NC (1995) Degradation of hexahydro-1,3,5-trinitro-1,3,5-triazine (RDX) by *Stenotrophomonas maltophilia* PB1. Appl Environ Microbiol 61:1318–1322.

Boopathy R (1994) Transformation of nitroaromatic compounds by methanogenic bacterium *Methanococcus* sp. (strain B). Arch Microbiol 162:131–137.

Boopathy R (2001) Enhanced biodegradation of cyclotetramethylenetetranitramine (HMX) under mixed electron-acceptor condition. Bioresour Technol 76:241–244.

Boopathy R, Manning JF (1996) Characterisation of partial anaerobic metabolic pathway for 2,4,6-trinitrotoluene degradation by sulfate-reducing bacterial consortium. Can J Microbiol 42:1203–1208.

Boopathy R, Melancon E (2004) Metabolism of compounds with nitro-functions by *Klebsiella pneumoniae* isolated from a regional wetland. Int Biodeterior Biodegrad 54:269–275.

Boopathy R, Kulpa CF, Wilson M (1993) Metabolism of 2,4,6-trinitrotoluene (TNT) by *Desulfovibrio* sp. (B strain). Appl Microbiol Biotechnol 39:270–275.

Boopathy R, Manning J, Kulpa CF (1997) Optimisation of environmental factors for the biological treatment of trinitrotoluene-contaminated soil. Arch Environ Contam Toxicol 32:94–98.

Boopathy R, Kulpa CF, Manning J (1998) Anaerobic biodegradation of explosives and related compounds by sulfate-reducing and methanogenic bacteria: a review. Bioresour Technol 63:81–89.

Brannon JM, Myers TE (1997) Review of fate and transport processes of explosives Technical Report IRRP-97-2. U.S Army Engineer Waterways Experiment Station, Vicksburg, MS.

Brannon JM, Pennington JC (2002) Environmental fate and transport process descriptors for explosives. ERDC/EL TR-02-10. U.S Army Engineer Research and Development Center, Vicksburg, MS.

Brannon JM, Price CB, Hayes C, Yost SL (2002) Aquifer soil cation substitution and adsorption of TNT, RDX, and HMX. Soil Sed Contam 11:327–338.

Bumpus JA, Tatarko M (1994) Biodegradation of 2,4,6-trinitrotoluene by *Phanerochaete chrysosporium*: identification of initial degradation products and the discovery of a TNT metabolite that inhibits lignin peroxidases. Curr Microbiol 28:185–190.

Burton DT, Turley SD, Peters GT (1994) The acute and chronic toxicity of hexahydro-1,3,5-trinitro-1,3,5-triazine (RDX) to the flathead minnow (*Pimephales promelas*). Chemosphere 29:567–579.

Caballero A, Lazaro JJ, Ramos JL, Esteve-Nunez A (2005a) PnrA, a new nitroreductase-family enzyme in the TNT-degrading strain *Pseudomonas putida* JLR11. Environ Microbiol 7:1211–1219.

Caballero A, Esteve-Nunez A, Zylstra GJ, Ramos JL (2005b) Assimilation of nitrogen from nitrite and trinitrotoluene in *Pseudomonas putida* JLR11. J Bacteriol 187:396–399.

Carr RS, Nipper M (2003) Assessment of environmental effects of ordnance compounds and their transformation products in coastal ecosystems. Technical Report TR-2234-ENV. Naval Facilities Engineering Service Center, Port Hueneme, CA.

Cataldo DA, Harvey SD, Fellows RM, Bean RM, McVeety BD (1989) Uptake of TNT by bush beans. Report ADA 223340. Pacific Northwest Laboratory, Richland, WA.

Coleman NV, Duxbury T (1999) Biodegradation of RDX by Rhodococcus sp. strain DN22 Presented at the 2nd International Symposium on the Biodegradation of Nitroaromatic Compounds and Explosives, Leesburg, VA. 8–9 September 1999.

Coleman NV, Nelson DR, Duxbury T (1998) Anaerobic biodegradation of hexahydro-1,3,5-trinitro-1,3,5-triazine (RDX) as a nitrogen source by a *Rhodococcus* sp. strain DN22. Soil Biol Biochem 30:1159–1167.

Coleman NV, Spain JC, Duxbury T (2002) Evidence that RDX biodegradation by *Rhodococcus* strain DN22 is plasmid-borne and involves a cytochrome p-450. J Appl Microbiol 93:463–472.

Conder JM, La Point TW, Bowen AT (2004a) Preliminary kinetics and metabolism of 2,4,6-trinitrotoluene and its reduced metabolites in an aquatic oligochaete. Aquat Toxicol 69:199–213.

Conder JM, La Point TW, Steevens JA, Lotufo GR (2004b) Recommendations for the assessment of TNT toxicity in sediment. Environ Toxicol Chem 23:141–149.

Davis JL, Wani AH, O'Neal BR, Hansen LD (2004) RDX biodegradation column study: comparison of electron donors for biologically induced reductive transformation in groundwater. J Hazard Mater 112:45–54.

Devlin JF, Klausen J, Schwarzenbach RP (1998) Kinetics of nitroaromatic reduction on granular iron in recirculating batch experiments. Environ Sci Technol 32:1941–1947.

Dilley JV, Tyson CA, Spanggord RJ, Sasmore DP, Newell GW, Dacre JC (1982) Short-term oral toxicity of 2,4,6-trinitrotoluene in mice, rats and dogs. J Toxicol Environ Health 9:565–585.

Dodard SG, Renoux AY, Hawari J, Ampleman G, Thiboutot S, Sunahara GI (1999) Ecotoxicity characterization of dinitrotoluenes and some of their reduced metabolites. Chemosphere 38:2071–2079.

Dodard SG, Renoux AY, Powlowski J, Sunahara GI (2003) Lethal and subchronic effects of 2,4,6-trinitrotoluene (TNT) on *Enchytraeus albidus* in spiked artificial soil. Ecotoxicol Environ Saf 54:131–138.

Dodard SG, Powlowski J, Sunahara GI (2004) Biotransformation of 2,4,6-trinitrotoluene (TNT) by enchytraeids (*Enchytraeus albidus*) *in vivo* and *in vitro*. Environ Pollut 131:263–273.

Dodard SG, Sunahara GI, Kuperman RG, Sarrazin M, Gong P, Ampleman G (2005) Survival and reproduction of enchytraeid worms, oligochaeta, in different soil types amended with energetic cyclic nitramines. Environ Toxicol Chem 24:2579–2587.

Donnelly KC, Chen JC, Huebner HJ, Brown KE, Autenrieth RL, Bonner JS (1997) Utility of four strains of white-rot fungi for the detoxification of 2,4,6-trinitrotoluene in liquid culture. Environ Toxicol Chem 16:1105–1110.

Drzyzga O, Gorontzy T, Schmidt A, Blotevogel KH (1995) Toxicity of explosives and related compounds to the luminescent bacterium *Vibrio fischeri* NRRL-B-11177. Arch Environ Contam Toxicol 28:229–235.

Dutta SK, Jackson MM, Hou LH, Powell D, Tatem HE (1998) Non-ligninolytic TNT mineralisation in contaminated soil by *Phanerochaete chrysosporium.* Bioremed J 2:97–103.

Eilers A, Rungeling E, Stundl UM, Gottschalk G (1999) Metabolism of 2,4,6-trinitrotoluene by the white-rot fungus *Bjerkandera adusta* DSM 3375 depends on cytochrome P-450. Appl Microbiol Biotechnol 53:75–80.

Ek H, Dave G, Sturve J, Almroth BC, Stephensen E, Forlin L, Birgersson G (2005) Tentative biomarkers for 2,4,6-trinitrotoluene (TNT) in fish (*Oncorhynchus mykiss*). Aquat Toxicol 72:221–230.

Emmrich M (1999) Kinetics of the alkaline hydrolysis of 2,4,6-trinitrotoluene in aqueous solution and highly contaminated soils. Environ Sci Technol 33:3802–3805.

Esteve-Nunez A, Ramos JL (1998) Metabolism of 2,4,6-trinitrotoluene by *Pseudomonas* sp. JLR11. Environ Sci Technol 32:3802–3808.
Fernando T, Aust SD (1991) Biodegradation of munition waste TNT (2,4,6-trinitrotoluene) and RDX (hexahydro-1,3,5-trinitro-1,3,5-triazine) by *Phanerochaete chrysosporium.* Ind Eng Chem ACS Symp 486:214–232.
Fernando T, Bumpus JA, Aust SD (1990) Bioremediation of TNT (2,4,6-trinitrotoluene) by *Phanerochaete chrysosporium.* Appl Environ Microbiol 56:1666–1671.
Fiorella PD, Spain JC (1997) Transformation of 2,4,6-trinitrotoluene by *Pseudomonas pseudoalcaligenes* JS52. Appl Environ Microbiol 63:2007–2015.
Fournier D, Halasz A, Spain J, Fiurasek P, Hawari J (2002) Determination of key metabolites during biodegradation of hexahydro-1,3,5-trinitro-1,3,5-triazine with *Rhodococcus* sp. strain DN22. Appl Environ Microbiol 68:166–172.
Fournier D, Halasz A, Spain J, Spanggord RJ, Bottaro JC, Hawari J (2004a) Biodegradation of the hexahydro-1,3,5-trinitro-1,3,5-triazine ring cleavage product 4-nitro-2,4-diazabutanal by *Phanerochaete chrysospotium.* Appl Environ Microbiol 70:1123–1128.
Fournier D, Halasz A, Thiboutot S, Ampleman G, Manno D, Hawari J (2004b) Biodegradation of octahydro-1,3,5,7-tetranitro-1,3,5,7-tetrazocine (HMX) by *Phanerochaete chrysosporium*: new insight into the degradation pathway. Environ Sci Technol 38:4130–4133.
French CE, Nicklin S, Bruce NC (1998) Aerobic degradation of 2,4,6-trinitrotoluene by *Enterobacter cloacae* PB2 and pentaerythritol tetranitrate reductase. Appl Environ Microbiol 64:2864–2868.
Frische T (2002) Screening for soil toxicity and mutagenicity using luminescent bacteria: a case study of the explosive 2,4,6-trinitrotoluene. Ecotoxicol Environ Saf 51:133–144.
Fuller ME, Manning JF Jr (1997) Aerobic gram-positive and gram-negative bacteria exhibit differential sensitivity to and transformation of 2,4,6-trinitrotoluene (TNT). Curr Microbiol 35:77–83.
Gensemer R, Caldwell R, Paulus S, Crawford P (2004) Derivation of toxicity reference values for the acute and chronic toxicity of RDX to marine organisms. PH022, Fourth SETAC World Congress, 14–18 November 2004, Portland, OR.
Glover DJ, Hoffsommer JC (1979) Photolysis of RDX. Identification and reactions of products. Technical Report NSWC TR-79-349. Naval Surface Weapons Centre, Silver Spring, MD.
Gong P, Wilke BM, Fleischmann S (1999) Soil-based phytotoxicity of 2,4,6-trinitrotoluene (TNT) to terrestrial higher plants. Arch Environ Contam Toxicol 36:152–157.
Gregory KB, Larese-Casanova P, Parkin GF, Scherer MM (2004) Abiotic transformation of hexahydro-1,3,5-trinitro-1,3,5-triazine by fell bound to magnetite. Environ Sci Technol 38:1408–1414.
Groom CA, Beaudet S, Halasz A, Paquet L, Hawari J (2001) Detection of the cyclic nitramine explosives hexahydro-1,3,5-trinitro-1,3,5-triazine (RDX) and octahydro-1,3,5,7-tetranitro-1,3,5,7-tetrazine (HMX) and their degradation products in soil environments. J Chromatogr A 909:53–60.
Haderlein SB, Weissmahr KW, Schwarzenbach RP (1996) Specific adsorption of nitroaromatic explosives and pesticides to clay minerals. Environ Sci Technol 30:612–622.

Haidour A, Ramos JL (1996) Identification of products resulting from biological reduction of 2,4,6-trinitrotoluene, 2,4-dinitrotoluene and 2,6-dinitrotoluene by *Pseudomonas* sp. Environ Sci Technol 30:2365–2370.

Hannink NK, Rosser SJ, Bruce NC (2002) Phytoremediation of explosives. Crit Rev Plant Sci 21:511–538.

Hawari J (2000) Biodegradation of RDX and HMX: from basic research to field application. In: Biodegradation of Nitroaromatic Compounds and Explosives. CRC Press, Boca Raton, FL. pp 277–310.

Hawari J, Halasz A, Beaudet S, Ampleman G, Thiboutot S (1999) Biotransformaiton of 2,4,6-trinitrotoluene (TNT) with *Phanerochaete chrysosporium* in agitated cultures at pH 4.5. Appl Environ Microbiol 65:2977–2986.

Hawari J, Beaudet S, Halasz A, Thiboutot S, Ampleman G (2000a) Microbial degradation of explosives: biotransformation versus mineralization. Appl Microbiol Biotechnol 54:605–618.

Hawari J, Shen CF, Guiot SR, Greer CW, Rho D, Sunahara G, Ampleman G, Thiboutot S (2000b) Bioremediation of highly energetic compounds: a search for remediation technologies. Water Sci Technol 42:385–393.

Hawari J, Halasz A, Sheremata T, Beaudet S, Groom C, Paquet L, Rhofir C, Ampleman G, Thiboutot S (2000c) Characterization of metabolites during biodegradation of hexahydro-1,3,5-trinitro-1,3,5-triazine (RDX) with municipal anaerobic sludge. Appl Environ Microbiol 66:2652–2657.

Hawari J, Halasz A, Beaudet S, Paquet L, Ampleman G, Thiboutot S (2001) Biotransformation routes of octahydro-1,3,5,7-tetranitro-1,3,5,7-tetrazocine by municipal anaerobic sludge. Environ Sci Technol 35:70–75.

Hawthorne SB, Lagadec AJM, Kalderis D, Lilke AV, Miller DJ (2000) Pilot-scale destruction of TNT, RDX, and HMX on contaminated soils using subcritical water. Environ Sci Technol 34:3224–3228.

Hess TF, Schrader PS (2002) Coupled abiotic-biotic mineralization of 2,4,6-trinitrotoluene (TNT). J Environ Qual 31:736–744.

Hodgson J, Rho G, Guiot SR, Ampleman G, Thinoutot S, Hawari J (2000) Tween 80 enhanced TNT mineralisation by *Phanerochaete chrysosporium.* Can J Microbiol 46:110–118.

Hofstetter TB, Heijman CG, Haderlein SB, Holliger C, Schwarzenbach RP (1999) Complete reduction of TNT and other (poly)nitroaromatic compounds under iron reducing subsurface conditions. Environ Sci Technol 33:1479–1487.

Huang G, Xiao H, Chi J, Shiu W, MacKay D (2000) Effects of pH on the aqueous solubility of selected chlorinated phenols. J Chem Eng Data 45:441–414.

Hughes JB, Wang C, Yesland K, Richardson A, Bhadra R, Bennett G, Rudolph F (1998) Bamberger rearrangement during TNT metabolism by *Clostridium acetobutylicum.* Environ Sci Technol 32:494–500.

Hwang S, Felt DR, Bouwer EJ, Brooks MC, Larson SL, Davis JL (2006) Remediation of RDX-contaminated water using alkaline hydrolysis. J Environ Eng ASCE 132:256–262.

Jarvis AS, McFarland VA, Honeycutt ME (1998) Assessment of the effectiveness of composting for the reduction of toxicity and mutagenicity of explosive-contaminated soil. Ecotoxicol Environ Saf 39:131–135.

Jenkins TF, Walsh ME, Thorne PG, Miyares PH, Ranney TA, Grant CL, Esparaza JR (1998) Site characterisation for explosives contamination at a military firing range impact area. Special Report CRREL 98-9. U.S Army Cold Regions Research and Engineering Laboratory, Hanover, NH.

Johnson LR, Davenport R, Balbach H, Schaeffer DJ (1994) Phototoxicology. 2. Near ultraviolet light enhancement of Microtox assays of trinitrotoluene and aminodinitrotoluenes. Ecotoxicol Environ Saf 27:23–33.

Johnson MS, Franke LS, Lee RB, Holladay SD (1999) Bioaccumulation of 2,4,6-trinitrotoluene and polychlorinated biphenyls through two routes of exposure in a terrestrial amphibian: is the dermal route significant? Environ Toxicol Chem 18:873–878.

Johnson MS, Holladay SD, Lippenholz KS, Jenkins LJ, McCain WC (2000) Effects of 2,4,6-trinitrotoluene in a holistic environmental exposure regime on a terrestrial salamander, *Ambystoma tigrinum*. Toxicol Pathol 28:334–341.

Johnson MS, Paulus HI, Salice CJ, Checkai RT, Simini M (2004) Toxicologic and histopathologic response of the terrestrial salamander *Plethodon cinereus* to soil exposure of 1,3,5-trinitrohexahydro-1,3,5-triazine. Arch Environ Contam Toxicol 47:496–501.

Johnson MS, Gogal RM, Larsen CT (2005) Food avoidance behavior to dietary octahydro-1,3,5,7-tetranitro-1,3,5,7-tetrazocine (HMX) exposure in the northern bobwhite (*Colinus virginianus*). J Toxicol Environ Health Part A Curr Issues 68:1349–1357.

Jones AM, Greer CW, Ampleman G, Thiboutot S, Lavigne J, Hawari J (1995) Biodegradability of selected highly energetic pollutants under aerobic conditions Presented at the 3rd Int Conf on In Situ and On Site Bioreclamation, San Diego CA. 24–27 April 1995.

Kalafut T, Wales ME, Rastogi VK, Naumova RP, Zaripova SK, Wild JR (1998) Biotransformation pattern of 2,4,6-trinitrotoluene by aerobic bacteria. Curr Microbiol 36:45–54.

Kaplan DL (1993) Biotechnology and bioremediaiton of organic energetic compounds. In: Marikas P (ed) Organic Energetic Compounds. Nova Science, New York, pp 373–416.

Kim HY, Song HG (2000a) Comparison of 2,4,6-trinitrotoluene degradation by seven strains of white rot fungi. Curr Microbiol 41:317–320.

Kim HY, Song HG (2000b) Transformation of 2,4,6-trinitrotoluene by white rot fungus *Irpex lacteus*. Biotechnol Lett 22:969–975.

Kim HY, Song HG (2003) Transformation and mineralization of 2,4,6-trinitrotoluene by the white rot fungus *Irpex lacteus*. Appl Microbiol Biotechnol 61:150–156.

Kim HY, Song HG (2005) Purification and characterization of NAD(P)H-dependent nitroreductase I from *Klebsiella* sp. C1 and enzymatic transformation of 2,4,6-trinitrotoluene. Appl Microbiol Biotechnol 68:766–773.

Kim HY, Bennett GN, Song HG (2002) Degradation of 2,4,6-trinitrotoluene by *Klebsiella* sp. isolated from activated sludge. Biotechnol Lett 24:2023–2028.

Kim J, Drew MC, Corapcioglu MY (2004) Uptake and phytotoxicity of TNT in onion plant. J Environ Sci Health Part A Toxic/Hazard Subst Environ Eng 39:803–819.

Kitts CL, Cunningham DP, Unkefer PJ (1994) Isolation of three hexahydro-1,3, 5-trinitro-1,3,5-triazine degrading species of the family *Enterobacteriaceae* from nitramine explosive-contaminated soil. Appl Environ Microbiol 60:4608–4611.

Kuperman RG, Checkai RT, Simini M, Phillips CT, Kolakowski JE, Kurnas CW, Sunahara GI (2003) Survival and reproduction of *Enchytraeus crypticus* (Oligochaeta, Enchytraeidae) in a natural sandy loam soil amended with the nitro-heterocyclic explosives RDX and HMX. Pedobiologia 47:651–656.

Kuperman RG, Checkai RT, Simini M, Phillips CT, Kolakowski JE, Kurnas CW (2005) Weathering and aging of 2,4,6-trinitrotoluene in soil increases toxicity to potworm *Enchytraeus crypticus*. Environ Toxicol Chem 24:2509–2518.

Kutty R, Bennett GN (2005) Biochemical characterization of trinitrotoluene transforming oxygen-insensitive nitroreductases from *Clostridium acetobutylicum* ATCC 824. Arch Microbiol 184:158–167.

Kutty R, Bennett G (2006) Studies on inhibition of transformation of 2,4,6-trinitrotoluene catalyzed by Fe-only hydrogenase from *Clostridium acetobutylicum*. J Ind Microbiol Biotechnol 33:368–376.

Labidi M, Ahmad D, Halasz A, Hawari J (2001) Biotransformation and partial mineralization of the explosive 2,4,6-trinitrotoluene (TNT) by rhizobia. Can J Microbiol 47:559–566.

Lachance B, Renoux AY, Sarrazin M, Hawari J, Sunahara GI (2004) Toxicity and bioaccumulation of reduced TNT metabolites in the earthworm *Eisenia andrei* exposed to amended forest soil. Chemosphere 55:1339–1348.

Lee SY, Brodman BW (2004) Biodegradation of 1,3,5-trinitro-1,3,5-triazine (RDX). J Environ Sci Health Part A Toxic/Hazard Subst Environ Eng 39:61–75.

Leggett DC (1985) Sorption of military explosive contaminants on bentonite drilling muds. CRREL Report 85-18. U.S Army Cold Regions Research and Engineering Laboratory, Hanover, NH.

Leggett DC, Jenkins TF, Murrman RP (1977) Composition of vapours evolved from military TNT as influenced by temperature, solid composition, age and source. CRREL Report 87-7. U.S Army Cold Regions Research and Engineering Laboratory, Hanover, NH.

Levine BS, Furedi EM, Gordon DE, Burns JM, Lish PM (1981) Thirteen-week toxicity study of hexahydro-1,3,5-trinitro-1,3,5-triazine (RDX) in Fischer 344 rats. Toxicol Lett 8:241–245.

Levine BS, Rust JH, Barkley JJ, Furedi EM, Lish PM (1990) Six month oral toxicity study of trinitrotoluene in beagle dogs. Toxicology 63:233–244.

Lewis A, Goszczynski T, Crawford RL, Korus RA, Admassu W (1996) Products of anaerobic 2,4,6-trinitrotoluene (TNT) transformation by *Clostridium bifermentans*. Appl Environ Microbiol 62:4669–4674.

Li H, Sheng GY, Chiou CT, Xu OY (2005) Relation of organic contaminant equilibrium sorption and kinetic uptake in plants. Environ Sci Technol 39:4864–4870.

Lotufo GR, Farrar JD (2005) Comparative and mixture sediment toxicity of trinitrotoluene and its major transformation products to a freshwater midge. Arch Environ Contam Toxicol 49:333–342.

Lotufo GR, Lydy MJ (2005) Comparative toxicokinetics of explosive compounds in sheepshead minnows. Arch Environ Contam Toxicol 49:206–214.

Lotufo GR, Farrar JD, Inouye LS, Bridges TS, Ringelberg DB (2001) Toxicity of sediment-associated nitroaromatic and cyclonitramine compounds to benthic invertebrates. Environ Toxicol Chem 20:1762–1771.

Lynch JC (2002) Dissolution kinetics of high explosive compounds (TNT, RDX, HMX) ERDC/EL TR-02-23. U.S Army Engineering Research and Development Center, Vicksburg, MS.

Lynch JC, Myers KF, Brannon JM, Delfino JJ (2001) Effects of pH and temperature on the aqueous solubility and dissolution rate of 2,4,6-trinitrotoluene (TNT), hexahydro-1,3,5-trinitro-1,3,5-triazine (RDX), and octahydro-1,3,5,7-tetranitro-1,3,5,7-tetrazocine (HMX). J Chem Eng Data 46:1549–1555.

Lynch JC, Brannon JM, Delfino JJ (2002) Dissolution rates of three high explosive compounds: TNT, RDX, and HMX. Chemosphere 47:725–734.

Mabey WR, Tse D, Baraze A, Mill T (1983) Photolysis of nitroaromatics in aquatic systems. Chemosphere 12:3–16.

McCormick NG, Feeherry FE, Levinson HS (1976) Microbial transformation of 2,4,6-trinitrotoluene and other nitroaromatic compounds. Appl Environ Microbiol 31:949–958.

McCormick NG, Cornell JH, Kaplan AM (1981) Biodegradation of hexohydro-1,3,5-trinitro-1,3,5-triazine. Appl Environ Microbiol 42:817–823.

McCormick NG, Cornell JH, Kaplan AM (1985) The anaerobic transformation of RDX and HMX and their acetylated derivatives. Technical Report A149464 (TR85–008). U.S Army Natick Research and Development Center, Natick, MA.

McGrath CJ (1995) Review of formulations for processes affecting the subsurface transport of explosives Technical Report IRRP-95-2. U.S. Army Corps of Engineers, Waterways Experiment Station, Vicksburg, MS.

Meyer SA, Marchand AJ, Hight JL, Roberts GH, Escalon LB, Inouye LS, MacMillan DK (2005) Up-and-down procedure (UDP) determinations of acute oral toxicity of nitroso degradation products of hexahydro-1,3,5-trinitro-1,3,5-triazine (RDX). J Appl Toxicol 25:427–434.

Michels J, Gottschalk G (1994) Inhibition of the lignin peroxidase of *Phanerochaete chyrsosporium* by hydroxyamino-dinitrotoluene, an early intermediate in the degradation of 2,4,6-trinitrotoluene. Appl Environ Microbiol 60:187–194.

MLA (1996) Surveys of the Beaufort's Dyke explosives disposal site. Fisheries Research Services Report No. 15/96. Marine Laboratory, Aberdeen, MD.

Monteil-Rivera F, Groom C, Hawari J (2003) Sorption and degradation of octahydro-1,3,5,7-tetranitro-1,3,5,7-tetrazocine in soil. Environ Sci Technol 37:3878–3884.

Montpas S, Samson J, Langois E, Lei J, Piche Y, Chenevert R (1997) Degradation of 2,4,6-trinitrotoluene by *Serratia marcescens*. Biotechnol Lett 19:291–294.

Morley MC, Shammas SN, Speitel GE (2002) Biodegradation of RDX and HMX mixtures: batch screening experiments and sequencing batch reactors. Environ Eng Sci 19:237–250.

Mukhi S, Pan X, Cobb, GP, Patino R (2005) Toxicity of hexahydro-1,3,5-trinitro-1,3,5-triazine to larval zebrafish (*Danio rerio*). Chemosphere 61:178–185.

NFESC (2000a) Toxicity of marine sediments and pore waters spiked with ordnance compounds Report CR 01-001-ENV. Naval Facilities Engineering Command, Port Hueneme, CA.

NFESC (2000b) Development of marine sediment toxicity for ordnance compounds and toxicity identification evaluation studies at selected naval facilities. Report CR 01-002-ENV. Naval Facilities Engineering Command, Port Hueneme, CA.

Nipper M, Carr RS, Biedenbach JM, Hooten RL, Miller K, Saepoff S (2001) Development of marine toxicity data for ordnance compounds. Arch Environ Contam Toxicol 41:308–318.

Ownby DR, Belden JB, Lotufo GR, Lydy MJ (2005) Accumulation of trinitrotoluene (TNT) in aquatic organisms. Part 1: Bioconcentration and distribution in channel catfish (*Ictalurus punctatus*). Chemosphere 58:1153–1159.

Park C, Kim TH, Kim S, Lee J, Kim SW (2002) Biokinetic parameter estimation for degradation of 2,4,6-trinitrotoluene (TNT) with *Pseudomonas putida* KP-T201. J Biosci Bioeng 94:57–61.

Park CW, Kim TH, Kim SY, Lee JW, Kim SW (2003a) Bioremediation of 2,4,6-trinitrotoluene contaminated soil in slurry and column reactors. J Biosci Bioeng 96:429–433.

Park C, Kim TH, Kim S, Kim SW, Lee J, Kim SH (2003b) Optimization for biodegradation of 2,4,6-trinitrotoluene (TNT) by *Pseudomonas putida*. J Biosci Bioeng 95:567–571.

Park J, Comfort SD, Shea PJ, Machacek TA (2004) Remediating munitions-contaminated soil with zerovalent iron and cationic surfactants. J Environ Qual 33:1305–1313.

Parris GE (1980) Covalent binding of aromatic amines to humates. 1. Reactions with carbonyls and quinones. Environ Sci Technol 14:1099–1106.

Pasti-Grigsby MB, Paszczynski A, Goszczynski S, Crawford DL, Crawford RL (1992) Influence of aromatic substitution patterns on azo dyes by *Streptomyces* sp. and *Phanerochaete chrysosporium*. Appl Environ Microbiol 58:3605–3613.

Pasti-Grigsby MB, Lewis TA, Crawford DL, Crawford RL (1996) Transformation of 2,4,6-trinitrotoluene (TNT) by actinomycetes isolated from TNT-contaminated and uncontaminated environments. Appl Environ Microbiol 62:1120–1123.

Pavlostathis SG, Jackson GH (2002) Biotransformation of 2,4,6-trinitrotoluene in a continuous-flow *Anabaena* sp. system. Water Res 36:1699–1706.

Pennington JC, Brannon JM (2002) Environmental fate of explosives. Thermochim Acta 384:163–172.

Pennington JC, Patrick WH Jr (1990) Adsorption and desorption of 2,4,6-trinitrotoluene by soil. J Environ Qual 19:559–567.

Peterson MM, Horst GL, Shea PJ, Comfort SD, Peterson RKD (1996) TNT and 4-amino-2,6-dinitrotoluene influence on germination and early seedling development of tall fescue. Environ Pollut 93:57–62.

Peterson MM, Horst GL, Shea PJ, Comfort SD (1998) Germination and seed development of switchgrass and smooth bromegrass exposed to 2,4,6-trinitrotoluene. Environ Pollut 99:53–59.

Phillips CT, Checkai RT, Wentsel RS (1993) Toxicity of selected munitions and munition-contaminated soil on the earthworm (*Eisenia foetida*). ERDEC-TR-037. Edgeworth Research Development and Engineering Center, U.S. Army Chemical and Biological Defence Agency, Aberdeen Proving Ground, MD.

Picka K, Friedl Z (2004) Phytotoxicity of some toluene nitroderivatives and products of their reduction. Fresenius Environ Bull 13:789–794.

Preuss AJ, Fimpel J, Diekert G (1993) Anaerobic transformation of 2,4,6-trinitrotoluene. Arch Microbiol 159:345–353.

Price CB, Brannon JM, Yost SL (1998) Transformation of RDX and HMX under controlled Eh/pH conditions Technical Report IRRP-98-2. U.S. Army Engineering Waterways Experimental Station, Vicksburg, MS.

Price RA, Pennington JC, Larson LS, Neumann D, Hayes CA (2002) Uptake of RDX and TNT by agronomic plants. Soil Sed Contam 11:307–326.

Pudge IB, Daugulis AJ, Dubois C (2003) The use of *Enterobacter cloacae* ATCC 43560 in the development of a two-phase partitioning bioreactor for the destruction of hexahydro-1,3,5-trinitro-1,3,5-*s*-triazine (RDX). J Biotechnol 100:65–75.
Reddy G, Chandra SAM, Lish JW, Qualls CW Jr (2000) Toxicity of 2,4,6-trinitrotoluene in Hispid Cotton rats (*Sigmodon hispidus*): haematological, biochemical and pathological effects. Int J Toxicol 19:169–177.
Renoux AY, Sarrazin M, Hawari J, Sunahara GI (2000) Transformation of 2,4,6-trinitrotoluene in soil in the presence of the earthworm *Eisenia andrei*. Environ Toxicol Chem 19:1473–1480.
Rho D, Hodgson J, Thiboutot S, Ampleman G, Hawari J (2001) Transformation of 2,4,6-trinitrotoluene (TNT) by immobilized *Phanerochaete chrysosporium* under fed-batch and continuous TNT feeding conditions. Biotechnol Bioeng 73:271–281.
Robidoux PY, Hawari J, Thiboutot S, Ampleman G, Sunahara GI (1999) Acute toxicity of 2,4,6-trinitrotoluene in earthworm (*Eisenia andrei*). Ecotoxicol Environ Saf 44:311–321.
Robidoux PY, Svendsen C, Caumartin J, Hawari J, Ampleman G, Thiboutot S, Weeks JM, Sunahara GI (2000) Chronic toxicity of energetic compounds in soil determined using the earthworm (*Eisenia andrei*) reproduction test. Environ Toxicol Chem 19:1764–1773.
Robidoux PY, Hawari J, Thiboutot S, Ampleman G, Sunahara GI (2001) Chronic toxicity of octahydro-1,3,5,7-tetranitro-1,3,5,7-tetrazocine (HMX) in soil determined using the earthworm (*Eisenia andrei*) reproduction test. Environ Pollut 111:283–292.
Robidoux PY, Hawari J, Bardai G, Paquet L, Ampleman G, Thiboutot S, Sudahara GI (2002a) TNT, RDX, and HMX decrease earthworm (*Eisenia andrei*) life-cycle responses in a spiked natural forest soil. Arch Environ Contam Toxicol 43:379–388.
Robidoux PY, Svendsen C, Sarrazin M, Hawari J, Thiboutot S, Ampleman G, Weeks JM, Sunahara GI (2002b) Evaluation of tissue and cellular biomarkers to assess 2,4,6-trinitrotoluene (TNT) exposure in earthworms: effects-based assessment in laboratory studies using *Eisenia andrei*. Biomarkers 7:306–321.
Robidoux PY, Bardai G, Paquet L, Ampleman G, Thiboutot S, Hawari J, Sunahara GI (2003) Phytotoxicity of 2,4,6-trinitrotoluene (TNT) and octahydro-1,3,5,7-tetranitro-1,3,5,7-tetrazocine (HMX) in spiked artificial and natural forest soils. Arch Environ Contam Toxicol 44:198–209.
Robidoux PY, Gong P, Sarrazin M, Bardai G, Paquet L, Hawari J, Dubois C, Sunahara GI (2004a) Toxicity assessment of contaminated soils from an antitank firing range. Ecotoxicol Environ Saf 58:300–313.
Robidoux PY, Dubois C, Hawari J, Sunahara GI (2004b) Assessment of soil toxicity from an antitank firing range using *Lumbricus terrestris* and *Eisenia andrei* in mesocosms and laboratory studies. Ecotoxicology 13:603–614.
Robidoux PY, Svendsen C, Sarrazin M, Thiboutot S, Ampleman G, Hawari J (2005) Assessment of a 2,4,6-trinitrotoluene-contaminated site using *Aporrectodea rosea* and *Eisenia andrei* in mesocosms. Arch Environ Contam Toxicol 48:56–67.
Rocheleau S, Kuperman RG, Martel M, Paquet L, Bardai G, Wong S (2006) Phytotoxicity of nitroaromatic energetic compounds freshly amended or weathered and aged in sandy loam soil. Chemosphere 62:545–558.

Rosen G, Lotufo GR (2005) Toxicity and fate of two munitions constituents in spiked sediment exposures with the marine amphipod *Eohaustorius estuarius*. Environ Toxicol Chem 24:2887–2897.

Saka M (2004) Developmental toxicity of *p,p′*-dichlorodiphenyltrichloroethane, 2,4,6-trinitrotoluene, their metabolites, and benzo[*a*]pyrene in *Xenopus laevis* embryos. Environ Toxicol Chem 23:1065–1073.

Samson J, Langlois E, Lei J, Piche Y, Chenevert R (1998) Removal of 2,4,6-trinitrotoluene and 2,4-dinitrotoluene by fungi (*Ceratocystis coerulescens, Lentinus lepideus* and *Trichoderma harzianum*). Biotechnol Lett 20:355–358.

Saupe A, Garvens HJ, Heinze L (1998) Alkaline hydrolysis of TNT and TNT in soil followed by thermal treatment of the hydrolysates. Chemosphere 36:1725–1744.

Schaefer M (2004) Assessing 2,4,6-trinitrotoluene (TNT)-contaminated soil using three different earthworm test methods. Ecotoxicol Environ Saf 57: 74–80.

Schafer R, Achazi RK (1999) The toxicity of soil samples containing TNT and other ammunition derived compounds in the enchytraeid and collembolan-biotest. Environ Sci Pollut Res 6:213–219.

Scheibner K, Hofrichter M, Herre A, Michels J, Fritsche W (1997) Screening for fungi intensively mineralising 2,4,6-trinitrotoluene. Appl Microbiol Biotechnol 47:452–457.

Scheidemann P, Klunk A, Sens C, Werner D (1998) Species dependent uptake and tolerance of nitroaromatic compounds by higher plants. J Plant Physiol 152: 242–247.

Seth-Smith HMB, Rosser SJ, Basran A, Travis ER, Dabbs ER, Nicklin S, Bruce NC (2002) Cloning, sequencing, and characterization of the hexahydro-1,3,5-trinitro-1,3,5-triazine degradation gene cluster from *Rhodococcus rhodochrous.* Appl Environ Microbiol 68:4764–4771.

Sherburne LA, Shrout JD, Alvarez PJJ (2005) Hexahydro-1,3,5-trinitro-1,3,5-triazine (RDX) degradation by *Acetobacterium paludosum*. Biodegradation 16: 539–547.

Sheremata TW, Hawari J (2000) Mineralization of RDX by the white rot fungus *Phanerochaete chrysosporium* to carbon dioxide and nitrous oxide. Environ Sci Technol 34:3384–3388.

Sheremata TW, Thiboutot S, Ampleman G, Paquet L, Halasz A, Hawari J (1999) Fate of 2,4,6-trinitrotoluene and its metabolites in natural and model soil systems. Environ Sci Technol 33:4002–4008.

Sheremata TW, Halasz A, Paquet L, Thiboutot S, Ampleman G, Hawari J (2001) The fate of the cyclic nitramine explosive RDX in natural soil. Environ Sci Technol 35:1037–1040.

Simini M, Wentsel RS, Checkai RT, Phillips CT, Chester NA, Major MA, Amos JC (1995) Evaluation of soil toxicity at Joliet Army Ammunition Plant. Environ Toxicol Chem 14:623–630.

Simini M, Checkai RT, Kuperman RG, Phillips CT, Kolakowski JE, Kurnas CW, Sunahara GI (2003) Reproduction and survival of *Eisenia fetida* in a sandy loam soil amended with the nitro-heterocyclic explosives RDX and HMX. Pedobiologia 47:657–662.

Singh J, Comfort SD, Hundal LS, Shea PJ (1998) Long-term RDX sorption and fate in soil. J Environ Qual 27:572–577.

Singh J, Comfort SD, Shea PJ (1999) Iron-mediated remediation of RDX-contaminated water and soil under controlled Eh/pH. Environ Sci Technol 33:1488–1494.

Smith JN, Pan XP, Gentles A, Smith EE, Cox SB, Cobb GE (2006) Reproductive effects of hexahydro-1,3,5-trinitroso-1,3,5-triazine in deer mice (*Peromyscus maniculatus*) during a controlled exposure study. Environ Toxicol Chem 25:446–451.

Spadaro JT, Gold MH, Renganathan V (1992) Degradation of azo dyes by the lignin degrading fungus *Phanerochaete chrysosporium*. Appl Environ Microbiol 58: 2397–2401.

Spanggord RJ, Mill T, Chou TW, Mabey WH, Smith JH, Lee S (1980a) Environmental fate studies on certain munition waste-water constituents. SRI Report LSU-7934. SRI International, Menlo Park, CA.

Spanggord RJ, Mill T, Chou TW, Mabey WH, Smith JH, Lee S (1980b) Environmental fate studies on certain munition waste water constituents: Part 2. SRI Report LSU-7934. SRI International, Menlo Park, CA.

Spanggord RJ, Mabey RW, Chow TW, Hayes DL, Alfernese PL, Tse DS, Mill T (1982) Environmental fate studies of HMX. SRI Project LSU-4412. SRI International, Menlo Park, CA.

Spiker JK, Crawford DL, Crawford RL (1992) Influence of 2,4,6-trinitrotoluene (TNT) concentration on the degradation of TNT inexplosives-contaminated soil by the white rot fungus *Phanerochaete chrysosporium*. Appl Environ Microbiol 58:3199–3202.

Stahl JD, Aust SD (1993) Metabolism and detoxification of TNT by *Phanerochaete chrysosporium*. Biochem Biophys Res Commun 192:477–482.

Steevens JA, Duke BM, Lotufo GR, Bridges TS (2002) Toxicity of the explosives 2,4,6-trinitrotoluene, hexahydro-1,3,5-trinitro-1,3,5-triazine and octahydro-1,3,5,7-tetranitro-1,3,5,7-tetrazocine in sediments to *Chironomus tentans* and *Hyalella azteca*: low-dose hormesis and high-dose mortality. Environ Toxicol Chem 21:1475–1482.

Sublette KL, Ganapathy EV, Schwartz S (1992) Degradation of munitions waste by *Phanerochaete chrysosporium*. Appl Biochem Biotechnol 34/35:709–723.

Sunahara GI, Dodard S, Sarrazin M, Paquet L, Ampleman G, Thiboutot S, Hawari J, Renoux AY (1998) Development of a soil extraction procedure for ecotoxicity characterization of energetic compounds. Ecotoxicol Environ Saf 39: 185–194.

Sunahara GI, Dodard S, Sarrazin M, Paquet L, Hawari J, Greer CW (1999) Ecotoxicological characterization of energetic substances using a soil extraction procedure. Ecotoxicol Environ Saf 43:138–148.

Tekoah Y, Abeliovich NA (1999) Participation of cytochrome p-450 in the biodegradation of RDX by a Rhodococcus strain 8–9 September 1999. Presented at the 2nd International Symposium on the Biodegradation of Nitroaromatic Compounds and Explosives, Leesburg, VA.

Thompson KT, Crocker FH, Fredrickson HL (2005) Mineralization of the cyclic nitramine explosive hexahydro-1,3,5-trinitro-1,3,5-triazine by *Gordonia* and *Williamsia* spp. Appl Environ Microbiol 71:8265–8272.

Thompson PL, Ramer LA, Schnoor JL (1998) Accumulation of TNT and its transformation products in the roots of poplar trees. Environ Sci Technol 32:975–980.

Thompson PL, Ramer LA, Schnoor JL (1999) Hexahydro-1,3,5-trinitro-1,3,5-triazine translocation in poplar trees. Environ Toxicol Chem 18:279–284.
Thorn KA, Thorne PG, Cox LG (2004) Alkaline hydrolysis/polymerization of 2,4,6-trinitrotoluene: characterization of products by C-13 and N-15 NMR. Environ Sci Technol 38:2224–2231.
U.S. Department of Defense (1983) Airblast attenuation in entranceways and other typical components of Structures Small-Scale tests data report 1. Report AD-A157 002. Cameron Station, Alexanderia, VA.
Van Aken B, Agathos SN (2001) Biodegradation of nitro-substituted explosives by white-rot fungi: a mechanistic approach. Adv Appl Microbiol 48:1–77.
Van Aken B, Skubisz K, Naveau H, Agathos SN (1997) Biodegradation of 2,4,6-trinitrotoluene by the white-rot basidiomycete *Phlebia radiata.* Biotechnol Lett 19:813–817.
Van Aken B, Hofrichter M, Scheibner K, Hatakka AI, Naveau H, Agathos SN (1999) Transformation and mineralization of 2,4,6-trinitrotoluene (TNT) by manganese peroxidase from the white-rot basidiomycete *Phlebia radiata.* Biodegradation 10:83–91.
Van Aken B, Yoon JM, Just CL, Schnoor JL (2004a) Metabolism and mineralization of hexahydro-1,3,5-trinitro-1,3,5-triazine inside poplar tissues (*Populus deltoids* × *nigra* DN-34). Environ Sci Technol 38:4572–4579.
Van Aken B, Yoon JM, Schnoor JL (2004b) Biodegradation of nitro-substituted explosives 2,4,6-trinitrotoluene, hexahydro-1,3,5-trinitro-1,3,5-triazine, an octa-hydro-1,3,5,7-tetranitro-1,3,5-tetrazocine by a phytosymbiotic *Methylobacterium* sp. associated with poplar tissues (*Populus deltoides* × *nigra* DN34). Appl Environ Microbiol 70:508–517.
Vanderberg LA, Perry JJ, Unkefer PJ (1995) Catabolism of 2,4,6-trinitrotoluene by *Mycobacterium vaccae.* Appl Microbiol Biotechnol 43:937–945.
Vila M, Pascal-Lorber S, Rathahao E, Debrauwer L, Canlet C, Laurent F (2005) Metabolism of [C-14]-2,4,6-trinitrotoluene in tobacco cell suspension cultures. Environ Sci Technol 39:663–672.
Vorbeck CV, Lenje H, Fischer P, Knackmuss H-J (1994) Identification of a hydride-Meisenheimer complex as a metabolite of 2,4,6-trinitrotoluene by *Mycobacterium* sp. J Bacteriol 176:932–934.
Walsh ME, Jenkins TF, Schnitker PS, Elwell JW, Stutz MH (1993) Evaluation of analytical requirements associated with sites potentially contaminated with residues of high explosives. CRREL Report 93-5. U.S. Army Cold Regions Research and Engineering Laboratory, Hanover, NH.
Walsh ME, Racine CH, Jenkins TF, Gelvin A, Ranney TA (2001) Sampling for explosives residues at Fort Greely, Alaska. ERDC/CRREL TR-01-15. U.S. Army Cold Regions Research and Engineering Laboratory, Hanover, NH.
Walsh ME, Hewitt AD, Walsh MR, Jenkins TF, Stark J, Gelvin A, Douglas TA, Perron N, Lambert D, Bailey R, Myers K (2004) Range characterisation studies at Donnelly Training Area, Alaska: 2001 and 2002. Report ERDC/CRREL TR-04-3. U.S. Army Cold Regions Research and Engineering Laboratory, Hanover NH.
Weber RWS, Ridderbusch DC, Anke H (2002) 2,4,6-Trinitrotoluene (TNT) tolerance and biotransformation potential of microfungi isolated from TNT-contaminated soil. Mycol Res 106:336–344.

Weiss M, Geyer R, Gunther T, Kastner M (2004a) Fate and stability of C-14-labeled 2,4,6-trinitrotoluene in contaminated soil following microbial bioremediation processes. Environ Toxicol Chem 23:2049–2060.

Weiss M, Geyer R, Russow R, Richnow HH, Kastner M (2004b) Fate and metabolism of N-15 2,4,6-trinitrotoluene in soil. Environ Toxicol Chem 23:1852–1860.

Weston (1993) Composting demonstration for explosives-contaminated soils at the Umatilla Depot Activity, Hermiston, Oregon. U.S. Army Environmental Center, Aberdeen, MD.

Xue SK, Iskandar IK, Selim HM (1995) Adsorption-desorption of 2,4,6-trinitrotoluene and hexahydro-1,3,5-trinitro-1,3,5-triazine in soils. Soil Sci 160:317–327.

Yamamoto H, Morley MC, Speitel GE, Clausen J (2004) Fate and transport of high explosives in a sandy soil: Adsorption and desorption. Soil Sed Contam 13:459–477.

Yin H, Wood TK, Smets BF (2005a) Reductive transformation of TNT by *Escherichia coli* resting cells: kinetic analysis. Appl Microbiol Biotechnol 69: 326–334.

Yin H, Wood TK, Smets BF (2005b) Reductive transformation of TNT by *Escherichia coli:* pathway description. Appl Microbiol Biotechnol 67:397–404.

Yinon J (1990) Toxicity and metabolism of explosives. CRC Press, Boca Raton, FL.

Yoon JM, Oh BT, Just CL, Schnoor JL (2002) Uptake and leaching of octahydro-1,3,5,7-tetranitro-1,3,5,7-tetrazocine by hybrid poplar trees. Environ Sci Technol 36:4649–4655.

Young DM, Unkefer PJ, Ogden KL (1997a) Biotransformation of hexahydro-1,3,5-trinitro-1,3,5-triazine (RDX) by a prospective consortium and its most effective isolate *Serratia marcescens*. Biotechnol Bioeng 53:515–522.

Young DM, Kitts CL, Unkefer PJ, Ogden KL (1997b) Biological breakdown of RDX in slurry phase reactors proceeds with multiple kinetically distinguishable paths. Biotechnol Bioeng 56:258–267.

Zeng K, Hwang HM, Zhang Y, Cook S (2004) Assessing cytotoxicity of photosensitized transformation products of 2,4,6-trinitrotoluene (TNT) and atrazine with freshwater microbial assemblages. Environ Toxicol 19:490–496.

Zhang CL, Hughes JB (2003) Biodegradation pathways of hexahydro-1,3,5-trinitro-1,3,5-triazine (RDX) by *Clostridium acetobutylicum* cell-free extract. Chemosphere 50:665–671.

Zhao HS, Paquet L, Halasz A, Manno D, Hawari M (2004) Metabolism of octahydro-1,3,5,7-tetranitro-1,3,5,7-tetrazocine by *Clostridium bifermentans* strain HAW-1 and several other H-2-producing fermentative anaerobic bacteria. FEMS Microbiol Lett 237:65–72.

Zhao JS, Halasz A, Paquet L, Beaulieu C, Hawari J (2002) Biodegradation of hexahydro-1,3,5-trinitro-1,3,5-triazine and its mononitroso derivative hexahydro-1-nitroso-3,5-dinitro-1,3,5-triazine by *Klebsiella pneumoniae* strain SCZ-1 isolated from an anaerobic sludge. Appl Environ Microbiol 68:5336–5341.

Zhao JS, Paquet L, Halasz A, Hawari J (2003a) Metabolism of hexahydro-1,3,5-trinitro-1,3,5-triazine through initial reduction to hexahydro-1-nitroso-3,5-dinitro-1,3,5-triazine followed by denitration in *Clostridium bifermentans* HAW-1. Appl Microbiol Biotechnol 63:187–193.

Zhao JS, Spain J, Hawari J (2003b) Phylogenetic and metabolic diversity of hexahydro-1,3,5-trinitro-1,3,5-triazine (RDX)-transforming bacteria in strictly anaerobic mixed cultures enriched on RDX as nitrogen source. FEMS Microbiol Ecol 46:189–196.

Zhao JS, Greer CW, Thiboutot S, Ampleman G, Hawari J (2004a) Biodegradation of the nitramine explosives hexahydro-1,3,5-trinitro-1,3,5-triazine and octahydro-1,3,5,7-tetranitro-1,3,5,7-tetrazocine in cold marine sediment under anaerobic and oligotrophic conditions. Can J Microbiol 50:91–96.

Zhao JS, Spain J, Thiboutot S, Ampleman G, Greer C, Hawari J (2004b) Phylogeny of cyclic nitramine-degrading psychrophilic bacteria in marine sediment and their potential role in the natural attenuation of explosives. FEMS Microbiol Ecol 49:349–357.

Zhao JS, Manno D, Beaulieu C, Paquet L, Hawari J (2005) *Shewanella sediminis* sp. nov., a novel Na^+-requiring and hexahydro-1,3,5-trinitro-1,3,5-trinitro-degrading bacterium from marine sediment. Int J Syst Evol Microbiol 55:1511–1520.

Zhao JS, Manno D, Leggiadro C, O'Neill D, Hawari J (2006) *Shewanella halifaxensis* sp. nov., a novel obligately respiratory and denitrifying psychrophile. Int J Syst Evol Microbiol 56:205–212.

Manuscript received September 9, accepted September 28, 2006.

Index

MIX
Papier aus verantwortungsvollen Quellen
Paper from responsible sources
FSC® C105338

If you have any concerns about our products,
you can contact us on
ProductSafety@springernature.com

In case Publisher is established outside the EU,
the EU authorized representative is:
Springer Nature Customer Service Center GmbH
Europaplatz 3, 69115 Heidelberg, Germany

Printed by Libri Plureos GmbH
in Hamburg, Germany